RATING PLACES

A Geographer's View on Quality of Life

Susan L. Cutter
Department of Geography
Rutgers University
New Brunswick
New Jersey

RESOURCE PUBLICATIONS
IN GEOGRAPHY

Library of Congress Card Number 85-13469
ISBN 0-89291-191-3

Library of Congress Cataloging in Publication Data

Cutter, Susan L.
    Rating places.

    (Resource publications in geography)
    Bibliography: p.
    1. Anthropo-geography — United States. 2. Quality
of life — United States. 3. Cities and towns — United
States. I. Title. II. Series.
GF503.C87 1985   307'.0973   85-13469
ISBN 0-89291-191-3

Publication Supported by the A.A.G.

Graphic Design by CGK

Printed by Commercial Printing Inc.
State College, Pennsylvania

Foreword

It is not surprising that some folks take seriously the ranks attributed to places. At a global scale, such indices as the Physical Quality of Life Index or the Index of Net Social Progress provide markers along the desired route to more pervasive and equitable development of nations. Social and environmental indicators provide measures that document progress of public policy initiatives and the emergence of further problems at regional and local levels. Among American cities, place boosterism is no idle pastime, but part of important industries that entice tourists, foster municipal growth and public investment, and attract industrial location decisions.

For geographers, *place* is central in our discipline, just as place is rooted in the very essence of being. Through time and across scholarly traditions we have understood places from a myriad of perspectives, finding a significant unity in seeing places through a synthesis of historical, cultural, ecological, and scientific understanding. However, the careful, thoughtful, sometimes hesitating understanding built by scholars is quickly overshadowed in the public mind and popular media by daring and controversial place-rating schemes. "In one man's mind, we're No. 79" (the mind belonging to SUNY-Cortland geographer Robert M. Pierce), instead of being 135th in the 1985 edition of the *Places Rated Almanac,* boasted a local newspaper in Lancaster, PA (Lancaster *Sunday News,* August 4, 1985, B1).

Nevertheless, we cannot take lightly the kinds of rating schemes that are popular today (but rooted in a half-century of forebears) nor dismiss them as unimportant. The City of Tulsa and its Metropolitan Tulsa Chamber of Commerce filed (but eventually dropped) a $26 million lawsuit against a researcher who admitted errors in ranking the city's quality of life. That alone should give serious pause to researchers who might view rating places as a trivial matter! In *Rating Places: A Geographer's View on Quality of Life,* author Susan Cutter puts trendy and popular place-rating studies in *their* place. After examining the challenges of understanding and measuring quality of life, Cutter looks at the broad realm of ranking and rating places. Her analysis of rating schemes points out problems of relying on existing data sources and ignoring important human perceptual dimensions of satisfaction with places. Given the complexity and difficulty of creating meaningful rating schemes, it is understandable that this book does not add yet another rating system — although it does draw on similar work by geographers such as Cutter, as well as studies by other investigators. However disappointed you might be not to find a scheme that rated your favorite haunts a bit higher than other rankings, this book will provide a timely and necessarily critical perspective on assessing quality of life.

Resource Publications in Geography are sponsored by the Association of American Geographers, a professional organization whose purpose is to advance studies in geography and to encourage the application of geographic research in education, government, and business. This series traces its origins to the Association's Commission on College Geography, whose *Resource Papers* were launched in 1968. Eventually 28 papers were published under sponsorship of the Commission through 1974 with the assistance of the National Science Foundation. Continued NSF support after completion of the Commission's work permitted the *Resource Papers for College Geography* to meet the original series goals for an additional four years and sixteen volumes:

The Resource Papers have been developed as expository documents for the use of both the student and the instructor. They are experimental in that they are designed to supplement existing texts and to fill a gap between significant research in American geography and readily accessible materials. The papers are concerned with important concepts or topics in modern geography and focus on one of three general themes: geographic theory, policy implications, or contemporary social relevance. They are designed to implement a variety of undergraduate college geography courses at the introductory and advanced level.

The popularity and usefulness of the two series suggested the importance of their continuation after 1978, once a self-supporting basis for their publication had been established.

For the **Resource Publications,** the original goals remain paramount. However, they have been broadened to include the continuing education of professional geographers as well as communication with the public on contemporary issues of geographic relevance. This monograph was developed, printed, and distributed under the auspices of the Association, whose members served in advisory and review roles during its preparation. The ideas presented, however, are the author's and do not imply AAG endorsement.

The author, editor, and Advisory Board hope that this book will enable geographers to contribute more meaningfully to the development of quality of life assessments and to the appropriate critique of popular investigations of this sort. We also hope that this book will enhance non-geographers' appreciation of the role of geography as a discipline in an important aspect of public policy and contemporary life. Fundamentally, we *all* are geographers when we consciously experience and understand places!

C. Gregory Knight, *The Pennsylvania State University*
Editor, Resource Publications in Geograpy

Resource Publications Advisory Board

James S. Gardner, *University of Waterloo*
Patricia Gober, *Arizona State University*
Charles M. Good, Jr., *Virginia Polytechnic Institute and State University*
Sam B. Hillard, *Louisiana State University*
Phillip C. Muehrcke, *University of Wisconsin, Madison*
Thomas J. Wilbanks, *Oak Ridge National Laboratory*

Preface and Acknowledgments

The quality of life in different places is subject to considerable debate. The popularity of quality of life studies is shown by best-selling books such as Boyer and Savageau's *Places Rated Almanac* (1981, 1985); Bowman, Giuliani and Minge's *Finding Your Best Place to Live in America* (1981), and Bayless's *The Best Towns in America* (1983). The fascination with comparing places to live, long a geographical preoccupation, is now reaching mass audiences. Pop quizzes appear in local newspapers, and *The National Enquirer* encourages you to check to see how your hometown rates in articles on the best and worst places to live in America.

Where is the "best" urban place to live in the United States? The "worst?" Depending on the survey, the "best" metropolitan areas may be Pittsburgh, Atlanta, or Portland, Oregon. The "worst" may be Yuba City, California; Newark, New Jersey; or the towns of Lawrence and Haverhill, Massachusetts. Many of these rankings are presented to the public as definitive and highly sophisticated statistical analyses, and their findings provoke controversy. Local news headlines such as "Worst-city Ranking Rankles Mayor," "Best, Worst Ratings Bring Cheers, Jeers," "All Riled Up About Ratings" and "Fresno's Dubious Distinction" are common. Local pride is wounded when a favorite place fails in a test using supposedly objective indicators. Media attention surrounding these rankings and reactions to them reflects the highly volatile nature of comparisons of quality of life among places. The interpretations of the ratings by the public range from acceptance as truth (by residents in highly rated places) through skepticism to moral outrage and indignation (by residents of lowest-ranked places).

While the study of quality of life is an inherently geographical endeavor, the geographic profession has made only minor contributions to the advancement of theory and practice in ranking places according to their quality of life. The purpose of this book is to provide an overview of the quality of life literature and to critique these efforts from a geographical perspective. A geographer's view is very different from the sociologist's or economist's: geographers are interested in not only social indicators of quality of life, but also in environmental and perceptual ones. In addition, our overriding concern with place forces us to evaluate *all* the attributes of places — social, environmental, perceptual — and then seek comparisons between places based on these attributes. Geographers believe in the uniqueness of places and are therefore reluctant to make value judgments concerning best or worst places, because every place is different and has both good and bad attributes.

Chapter 1 defines the concept of quality of life and reviews the evolution of an integrated approach to quality of life research. Chapter 2 describes measurement techniques and the role of quality of life indicators in distinguishing areas. Chapter 3 critiques empirical studies of quality of life which rank places at a variety of spatial scales. Finally, Chapter 4 describes the reasons for and public interest in quality of life.

vi   *Preface and Acknowledgements*

The final research and writing of this publication was partially supported by the Rutgers University Research Council's FASP Program. Cartographic assistance was provided by the Cartography Lab at the Rutgers Center for Coastal and Environmental Studies. There are also a number of individuals who have contributed either directly or indirectly to the inception, development and actual production of this *Resource Publication*. The manuscript was improved enormously by critical comments and selected barbs from George Carey and Frank Popper (Middlesex-Somerset-Hunterdon, NJ, Rank 29). In addition, Greg Knight (State College, PA, Rank 213) provided invaluable editorial guidance and strong words of encouragement. The maps and illustrations were done by Gina Bedoya and Michele Pronio (also from Rank 29). In addition to these helpful folks, a number of other people indirectly contributed to this effort by influencing my own thinking on the subject. My parents (Stockton, CA, Rank 282) contributed to my initial interest in places and geography by their constant relocations during my formative years starting in 17, then to unranked, back to 17, then to 5, 97, and finally 26. Brian J.L. Berry (now in Pittsburgh, PA, Rank 1) introduced me to the formal study of quality of life indicators during my graduate years at the University of Chicago (Rank 26), which is where this interest in quality of life was first developed. Caleb Warner (Boston, MA, Rank 2) steadfastly argued that ratings had a bit of whimsy to them, because Lawrence-Haverhill, MA (last in 1981) was not such a bad place and certainly was not the worst place in the country. He was right; it moved to 154 in 1985. Finally, I like to especially thank Hilary and Bill Renwick and Bob and Lynn Roundy (Middlesex-Somerset-Hunterdon, NJ, Rank 29) who have improved my own individual quality of life over the last six years. While these individuals have contributed greatly to this publication, any errors or omissions are, of course, my own responsibility.

Susan L. Cutter
Middlesex-Somerset-Hunterdon, NJ.
(Rank 29 in the 1985 edition of *Places Rated Almanac*)

Contents

List of Figures

List of Tables

1

Defining Quality of Life

Quality of life: it is both a perennial concern . . . and a rather recent preoccupation. It is a topic that permits many different definitions, and for which there is no widely agreed-upon index which allows us to monitor changes in that quality. — Report of the President's Commission on The Quality of American Life in the Eighties

Places are unique. They arouse special feelings ranging from topophilia, or love of place, to unexplained attachments to place. Geographers are keenly aware of these feelings about, and attachments to, place. If offered a choice, few people would want to live in a place that others rate as the worst city, state, or country. A city's rating or reputation, for example, could prompt moves to a different city with a 'better' quality of life. Without such a choice, residents often make subtle adaptations to places with the 'worst' quality of life. They acknowledge negative aspects of the area, but remind others that they are stuck there. They may completely ignore popularized rating guides, or they rationalize the poor ratings, suggesting instead that the person or study commenting on their city, state, or country simply does not know any better. Before we can examine these conflicting and often controversial notions of place, we must first answer the following questions: What does quality of life mean, and what constitutes a good or bad quality of life in a particular place?

A General Definition

Quality of life is broadly defined as an individual's happiness or satisfaction with life and environment including needs and desires, aspirations, lifestyle preferences, and other tangible and intangible factors which determine overall well-being. When an individual's quality of life is aggregated to the community level, the concept is linked to existing social and environmental conditions such as economic activity, climate, or the quality of cultural institutions. It includes both tangible and intangible measures reflecting local consensus on the community's values and goals.

A geographical definition of quality of life incorporates the concept of individual well-being but focuses more on *places* rather than individuals. As with an individual's quality of life, the geographic definition includes both objective and subjective measures of social and environmental conditions in a place and how these conditions are experienced by the people living there. The art — science to some — of evaluating the quality of life in a place involves two *states:* a *goal state* and an *appraisal state.* A goal state is a collective impression, a desired environment — living conditions we would like to have, or to which we aspire. This goal state is subjective, culturally biased, and based upon collective images of what a place *ought* to be. The appraisal state measures the actual environment — what *is* actually there. The appraisal state assumes some form of multivariate quantitative or qualitative criteria which are culturally defined and objective or subjective. The geographic definition of quality of life, therefore, is a measure of the difference between the goal state and the appraisal state, or the distance between the 'ought' environment and the 'is' environment. From a more practical standpoint, geographical quality of life is the measurement of the conditions of place, how those conditions are experienced and evaluated by individuals, and the relative importance of each of these to the individual.

When applied to places rather than individuals, quality of life research requires the analysis of both objective conditions and more subjective assessments of these conditions among and between places. Tangible or intangible quality of life criteria are a major component of locational decisions by individuals and institutions such as migration or siting a new business or industry. Faced with a complex locational problem, decision makers all too often reduce measurement of quality of life to a simple statistical measure reflecting comparative judgments over time about the welfare of society or conditions in a particular city, state, or country. A similar approach has been adopted at the community level, where quality of life measures are used as a decision making aid in evaluating the effectiveness of social and environmental programs and in allocating program funds. Quality of life measures have also been used at the national and international level. Here, states, regions and countries are ranked or evaluated from best to worst using similar measurement techniques, despite a major change in scale.

A Conceptual Model

As a general concept, quality of life incorporates objective and subjective information along a number of dimensions. From our geographical perspective, with the emphasis on place, three dimensions are important: social, environmental, and perceptual. Each dimension includes criteria that range from totally objective to highly subjective and each dimension is related to the other. These three aspects of a geographic measurement of quality of life are presented in a conceptual model (Figure 1).

One must not only consider the objective conditions of the social environment (crime, housing, income) in evaluating places, but the physical environment (climate, pollution, recreation) as well. In addition, some measure of individual satisfaction with, or subjective assessment of, these conditions is needed. The third component, perceptual, incorporates the relative importance of objective conditions (social and environmental) with a more subjective assessment of people's image of a place, their views toward that place, and their experiences and attachments to a place. Often, our individual image of a place overtly or covertly biases our evaluation of the quality of life found there. These perceptual indicators of human attachment to place are as important to quality of life from a geographical perspective as are the more traditional measures of existing social and environmental conditions. By their very nature, perceptual indicators are hard to quantify and can unintentionally color an apparently objective methodology of rating places.

Quality of life as a concept incorporates the relationship between all three model elements. By virtue of training and interests, geographers have a keen understanding

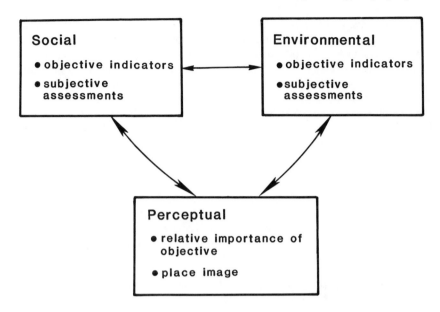

FIGURE 1 COMPONENTS OF QUALITY OF LIFE.

of each of these elements, resulting in a holistic view of human-environment systems. In this respect, geographers are particularly well-suited to conduct and evaluate quality of life analyses. Most rankings of places (and the indicators underlying them) are not explicit and contain hidden methodological flaws. Utilizing spatial data analysis techniques, geographers can make these explicit, drawing conclusions from them, particularly about the reputations of places. This is the primary contribution of a geographical perspective to quality of life.

Public Policy and Quality of Life: A Review

The term "quality of life" did not receive widespread usage until the 1960s-1970s, although the concept, embodied in such terms as "general welfare," "social accounting," and "social indicators," was used much earlier. One of the first to publicize the term was James Michener who wrote a short monograph in 1970 titled *A Quality of Life*. The monograph described his personal reflections on the quality of American life over the last 200 years, including the strengths of American society (the Constitution, education, economic abundance, and religion) and the problems it faces (rapid social and environmental change, increasing population). In addition, Michener detailed his views on the continuing prospects for this good American life during the rest of the century.

The development of standardized quality of life indicators began in the 1950s. After World War II there was a growing need to measure the social welfare of the country based on more than simple economic indicators. In response to the need for a broad description of trends in American life, the social science community began compiling data on national trends. The measures they developed were used to evaluate not only the nation's economic health but the non-economic facets of American life as well. While providing a purely descriptive statistical accounting of societal well-being, this monitoring effort did not offer any conclusions about the causes, effects, or meanings of social conditions or social prosperity.

During the 1960s, a number of ambitious governmental programs were established to provide more statistical information on social trends and societal well-being. This approach was endorsed by such academic bodies as the Life Science Panel of the President's Science Advisory Committee (1962), the Russell Sage Foundation (1965), the National Academy of Sciences (1969) and the Social Science Research Council (1969). The purpose of these programs was to acquire a uniform, national statistical data base on existing social conditions and to monitor the rate and direction of change in these social conditions. These "social indicators" were intended as quantitative measures of social conditions and trends, thus providing conclusions about the effects of various governmental actions. They were to serve many goals, ranging from measuring existing social conditions, to forecasting future social trends and to social engineering. Soon after the social indicators program was established, policy makers and social scientists recognized that it would not fulfill all of the prescribed goals, and, in fact, the only intent that would be realized was the measurement of existing social conditions. The task of the social indicators program was so large that it was virtually doomed to failure from the start, due in part to problems with quantifying abstract concepts such as social well-being, but also as a result of problems in summarizing these measures into a framework that could usefully define quality of life.

Shortly after the development of social indicators, researchers and policy makers began working on environmental indices. These environmental measures were designed to judge the success of environmental monitoring programs and to assess the effectiveness of the regulatory programs mandated by the environmental quality legislation passed in the 1970s. The use of indices provided a means to reduce the large quantity of data to a simpler form. In this context, a single *indicator* is derived from one variable such as the number of days an area exceeds the ambient SO_2 standard. An *index* provides a composite single value for an environmental component (air quality) which is derived from 2 or more indicators (Ott 1978).

Environmental indicators and indices have been used to assist environmental managers in allocating funds, determining policy priorities, enforcing environmental quality standards, establishing trends, and analyzing environmental quality for the public and scientific community (Thomas 1972; Inhaber 1976). Perhaps the best example of the current public use of environmental indicators is the issuance of daily air pollution levels. The primary use of environmental quality indices, however, has been to monitor the effectiveness of regulatory programs.

Many social scientists, including geographers and psychologists, were distressed by the sole use of objective measures of social conditions and environmental quality. They argued for the need to include subjective evaluations of environmental and social conditions as well. The need to develop and test appropriate methods to measure these subjective evaluations provided a new stimulus to the fields of environmental psychology, behavioral geography, and sociology.

Throughout the late 1960s and 1970s, researchers in many disciplines were busily engaged in studies of various aspects of quality of life. Unfortunately, very few of them went beyond their own disciplinary biases to incorporate findings from other fields. The federal government was keenly aware of the problem and attempted to include all aspects of social, environmental and perceptual well-being under the general heading of quality of life (USEPA 1973a, 1973b). Using the talents of academic researchers in the fields of sociology, economics, political science, psychology, environmental science, and geography, the federal government was interested in a comprehensive quantitative measure of quality of life that included not only social data, but environmental and perceptual data as well. As a discipline, geography was particularly well-suited to move in this direction, yet geographers were rather slow in entering the field.

The Geographer's Role

Place is a fundamental concern for all geographers regardless of training and field of interest. As a result, geographers approach the concept of place and the evaluation of places from many different points of view. The lack of geographical input into quality of life research may be attributable to the myriad of ways in which we approach the subject of place and the evaluation of places.

Although the phrase, "quality of life," is relatively new, from a spatial perspective the concept has intellectual antecedents in the descriptive place studies done by early geographers such as Strabo and by later geographers working in the regional or area-studies tradition such as Carl Sauer and Paul Vidal de la Blache (see James 1972; Lukermann 1964; and Relph 1976 for a review of the role of place in geographic thought).

While important, place studies are only one facet of quality of life research. Certainly, our image of place and attachment to place influence our evaluation of places, our behavior towards them, and ultimately the quality of life we find there (see, for example, the classic studies of Lynch 1960, 1972; Boulding 1961). The importance of our image of place is also emphasized by geographers who study place from a humanistic or phenomenological perspective. These geographers do not even attempt to evaluate the attributes of a place, much less rank them. Instead, they argue from an existential point of view that places are unique, have different meanings to different people, and are experienced in their own individual way. The most extensive review of these humanistic approaches to place is found in *Place and Placelessness* (1976) where Relph not only describes the essence, sense, and identity of places, but also the role of phenomenology in the study of place. There are also other geographers who have contributed to these humanistic perspectives on place (Tuan 1974, 1977; Buttimer and Seamon 1980; Relph 1981; Jackson 1984).

Social indicators research (the closest cousin to quality of life studies) has been approached by social scientists working in a normative mode with logical positivist perspectives. Geographers working in this field attempt to quantify the attributes of places and rank these according to predetermined assumptions or criteria. The field of social area analysis and factorial ecology is an outgrowth of this perspective (Shevky and Bell 1955; Hawley and Duncan 1957; Rees 1972; Johnston 1976).

As defined by our conceptual model, geographic research on quality of life incorporates elements from both the humanistic and positivist perspectives. However, very few studies which could legitimately be called quality of life are conducted by geographers. A geographer working in the positivist mode would be reluctant to quantify the image or attachment to place as a variable in his analysis. Conversely, any attempt at quantifying human experience is the antithesis of the humanistic approach.

Although rating places has broad popular appeal, interest in and techniques for systematically measuring quality of life were only recently addressed by the geographic community. Whereas Smith and colleagues (1973) examined place comparisons of social well-being in the early 1970s, and Hill and others (1973) described general conditions throughout the country, it was not until 1981 that formal papers using the term "quality of life" were presented at the national meetings of the Association of American Geographers. Three years later, a paper presented at the 1984 annual meetings (Pierce 1984) generated a record 400 calls to the convention hotel from the media. Quality of life as an explicit concept is now so widespread in its use within the profession that it can be found in both introductory geographical texts (Hartshorn 1981; Norris *et al.* 1982) and advanced urban and social geography texts (Ley 1982). Moreover, it was the topic of an Association of American Geographers presidential address (Helburn 1982).

Geography's contribution to the study of quality of life is our holistic perspective and our long history of creating areal data collection and analysis methods. This ability enables us to examine all the attributes of place including objective measures of existing conditions and subjective assessments of those conditions and their importance to individuals. As we shall see in the following chapters, not everyone shares this outlook. In fact, this is one of the major deficiencies of current research on quality of life. There is considerable work on the measurement of quality of life, but few studies systematically incorporate this holistic approach. Furthermore, very little of the quality of life research to date is used to describe places or distinguish them from one another, primary goals of geographers.

2

Measuring Quality of Life

> *Years ago I decided that even though I was free to live anywhere in the*
> *world, I would stick with rural Bucks County [PA], for I had seen nothing*
> *in my travels which surpassed it in its simple combination of natural*
> *beauty, orderliness and nearness to the big cities that I have enjoyed so*
> *much.* — *James Michener,* A Quality of Life *(1970)*

If a simple definition of quality of life is difficult to achieve, then universal agreement on how to measure it is virtually impossible. All too often, the variables chosen to measure quality of life are selected simply on the basis of available statistics rather than from any conceptual or theoretical consideration. Consequently, replication of studies is difficult, and comparability among them is virtually non-existent. Before we can critique such studies, however, we must first understand the methods, models, and data sources used in quality of life studies. This chapter examines the measurement of social, environmental, and perceptual indicators and comments on the strength, weakness, and limitation of each in determining the overall quality of life of a place. In addition, the use of indicators to differentiate and distinguish places is presented.

Models, Methods, and Data Sources

Social Indicators

During the 1960s, there was growing dissatisfaction among academics on the quality and quantity of social information available to public policy decision makers. In response to this, sociologists initiated a new research program aimed at improving social reporting in order to help decision makers make more informed judgments about social policy. This new research field was labeled "social indicators" (Bauer 1969; Gross 1969; Duncan 1969a, 1969b). During the 1970s, social indicators was a thriving research area within sociology. Volumes were written on the definition and meaning of social indicators, including theoretical and methodological issues (Sheldon and Moore 1968; Shonfield and Shaw 1972; Land and Spilerman 1975; Carley 1981; Land 1983; Duncan 1984) and the use of social indicators in public policy (Rossi and Gilmartin 1980; Smith 1981).

Despite this voluminous literature, there is still some disagreement over the measurement, value, and rationale for social indicators in formulating social policy. There is, however, a general consensus emerging on two aspects of their actual use (Rossi and Gilmartin 1980; Carley 1981; Land 1983). The first emphasizes the role of social indicators in policy formation. The second use focuses on the role of social indicators in social reporting and the need for social data. In the last decade, for example, social indicators were used primarily in three ways. They were used to develop a national data base which described current social conditions. These statistics not only described the current state of society (such as crime rate, unemployment, educational achievement) but were also used to evaluate the cost-effectiveness of public-service delivery systems such as health care, community services, and recreation. The most extensive practical use of social indicators, though, was in urban analysis. Social indicators were used at the intra-urban level to identify, describe, and classify social areas of the city. These urban-level social indicators are widely used in the United Kingdom, where community profiles and neighborhood surveys are often used to determine resource allocation to selective priority areas or specific population subgroups (Carley 1981). In the United States, social indicators were also used to compare and contrast different cities in terms of their social well-being. We will discuss these urban applications later in this chapter.

Although widely used, there are a number of methodological issues which continue to cloud social indicator research. Many times, social indicators involve a non-quantifiable or vague concept (such as good health, security) which is measured by some surrogate measure (life expectancy, crime rate). The surrogate measure is often chosen on the basis of data availability or ease of measurement rather than any theoretical consideration. More often than not, there is very little correlation between the quantified measure and non-quantified concept. Is living to age 75 (the measure of life expectancy) equivalent to good health? There are many other factors which indicate good health besides life expectancy, such as personal habits (eating, smoking, drinking), occupation, and so on.

A second methodological problem concerns non-theoretical and haphazard attempts at social indicators research, which makes comparing or replicating studies difficult. Problems abound, ranging from operationalizing concepts (such as socioeconomic status), to time parameters (using data from different years in the same study), to the spatial unit of analysis (neighborhood, census tract, city). A number of examples will help illustrate this point. While the researcher might be interested in the neighborhood level for 1985, data are only available at a census tract level for 1980. This is a function of how, when, and where the U.S. Bureau of the Census collects social information. Very often, researchers are limited in their research design because of their dependence on government data sources. As a result, studies may not measure what was originally intended or are not replicated because a statistic available in one area is not available in another. This lack of replication severely limits the use of social indicator research.

Finally, there are problems with variable selection, as well as with summarizing these variables into an overall measure of quality of life. One researcher may define socioeconomic status as father's occupation, while another defines it as educational level (number of years of school completed) *and* occupation (white or blue collar). Some studies utilize factor analysis to differentiate quality of life dimensions, making no assumptions about underlying trends. Others simply sum all the relevant statistics to create a composite or overall measure of well-being. The type of statistical technique used to analyze the data influences the results. The conclusions drawn using one type of technique (such as factor analysis) may be drastically different than those produced by a simple averaging of all indicators. As we shall see in Chapter 3, the lack of methodological rigor and appropriate statistical techniques often leads to erroneous and conflicting conclusions about quality of life.

Putting these criticisms aside, the availability, content, and sources of social data have steadily improved. One of the first systematic attempts to define the subject areas of social indicators in order to avoid confusion and vagueness in terminology and variable selection was made by Sheldon and Land (1972). They proposed five general topical areas ranging from socioeconomic welfare, to use of time, to consumption behavior (Table 1). Since then, the number of indicator variables has ranged anywhere from 4 in two subject areas to hundreds of variables in as many as 30 subject areas. There is no specific selection criteria for including variables in social indicators analyses, and oftentimes they appear to be related only to the interests of the governmental agencies or researchers involved.

Along with increasing sophistication and improvements in social indicators research, there was a concommitant rise in the number of available statistical publications containing social data. The most comprehensive are the products of the social indicators reporting authorized by the government (U.S. Bureau of the Census 1974, 1977, 1980). These *Social Indicators* volumes provide aggregated data for the U.S. as a whole. There are no spatial categories, so these volumes are relatively useless in comparing social conditions between places. A listing of other social indicators data sources including the number of variables, content (or subject areas), spatial coverage, and number of places covered is found in Table 2. Unfortunately, most of these publications provide data either on a national or international scale, not by neighborhood, city, or even region. Some, like the *Social Indicators* volumes, have no spatial reference at all. The U.S. Bureau of the Census, perhaps the best source of data on

TABLE 1   CONTENT AREAS OF SOCIAL INDICATORS

I   **Socioeconomic Welfare**
1. Population (composition, growth, distribution)
2. Labor force and employment
3. Income
4. Knowledge and technology
5. Education
6. Health
7. Leisure
8. Public safety and legal justice
9. Housing
10. Transportation
11. Physical environment
12. Social mobility and stratification

II   **Social Participation and Alienation**
1. Family
2. Religion
3. Politics
4. Voluntary Associations
5. Alienation

III   **Use of Time**
IV   **Consumption Behavior**
V   **Aspirations, Satisfactions, Acceptance, Morals**

Source: Sheldon and Land 1972.

TABLE 2   SELECTED SOURCES OF OBJECTIVE SOCIAL INDICATOR DATA[a]

Source	Number of Variables	Content Areas	Areal Coverage	Number of Places
Taylor and Jodice 1983	14	govt. size, resource allocation	nations	154
	9	population		
	10	wealth, production, size		
	8	inequality and well-being		
	13	social mobilization		
	9	economic structure		
	9	changes within countries		
	72	Total		
Garwood 1984	18	demographics	U.S. cities	199
	8	geography and climate		
	6	economic indicators		
	13	social indicators		
	14	cultural indicators		
	59	Total		
Kurian 1979, 1984	7	geography, climate	nations	190
	6	vital statistics/demography		
	18	population dynamics/family		
	4	race and religion		
	7	politics		
	6	foreign aid		
	8	defense		
	19	economics		
	11	finance and banking		
	12	trade		
	38	agricultural products		
	24	industry and mining		
	14	energy		
	18	labor		
	22	transportation/communication		
	18	consumption		
	10	housing		
	24	health and food		
	23	education		
	10	crime		
	13	media		
	5	culture and sports		
	317	Total		
U.S. Bureau of the Census, Annual Housing Survey, yearly	4	occupancy char.	SMSA/ regional/ central city/ rural	20
	22	characteristics of occupied units		
	6	evaluation of house/neighborhood		
	2	vacancy char.		
	15	char. of vacant units		
	49	Total		

TABLE 2   (continued)

Source	Number of Variables	Content Areas	Areal Coverage	Number of Places
U.S. Bureau	12	population	U.S. (aggregate)	1
of the Census	17	family		
1974, 1977,	14	housing		
1980	23	social security/welfare		
	36	health and nutrition		
	22	public safety		
	27	education and training		
	31	work		
	44	income, wealth, expenditures		
	19	culture, leisure, use of time		
	18	social mobility and participation		
	39	public perceptions		
	43	international comparisons		
	335	Total		
Abler and	2	physical attributes	SMSA/	20
Adams 1976	4	housing	Daily Urban	
	10	population composition	Systems	
	4	socioeconomic		
	3	special topics		
	21	urban problems		
	44	Total		
Boyer and	6	climate and terrain	U.S. Metro	277/
Savageau	3	housing	Areas	329
1981, 1985	10	health care & environment		
	2	crime		
	5	transportation		
	3	education		
	11	recreation		
	9	arts		
	4	economic		
	53	Total		

[a]In addition, Kurian (1983) provided data on 2,492 cities and towns and 3,142 counties; Marlin and Avery (1983) ranked cities in 267 different categories.

social conditions and trends, provides information on the block, tract, metropolitan, county and state level in both its population reports and annual housing surveys. The data are somewhat limited; they do not cover every aspect of social well-being for every scale of analysis and often must be supplemented with data from other federal, state, and local agencies.

Environmental Indicators

Following on the heels of the social indicators movement, environmental quality (EQ) indices were first developed by environmental and social scientists with strong financial encouragement from the U.S. Environmental Protection Agency. The indices fall into two main categories — "acceptable level" and "damage functions." The acceptable-level approach compares ambient levels of pollution, for example, to some pre-determined acceptable standard. Air and water quality, for instance, are compared to national primary and secondary ambient standards for selected pollutants. If an area exceeds the primary standard, enforcement ensues until the area complies.

The second type of index quantifies the potential for damage. Damage functions can either be related to dose-effect or economic costs. For example, a dose-effect approach statistically estimates the number of additional deaths in a geographic locale as a function of the measured concentration of each air pollutant, or it takes these estimates and translates them into an estimated death rate. This approach is widely used to assess the health effects of air pollution (Lave and Seskin 1977). Similarly, the economic impacts of air pollution are evaluated by quantifying the estimated relationship between exposure to pollutants and the various effects of pollution on materials, visibility, aesthetics, and human health (Bednarz 1975).

While there were many attempts to develop an overall environmental quality index, few, if any, met with success. Environmental quality data are not only subject to areal variations, but also change by season, time of day, and prevailing weather conditions. These data are extremely difficult to standardize and then summarize into a meaningful measure of total quality. Consequently, research focuses more on specific pollution indices, such as air quality or water quality, rather than composite indices of total environmental quality for a particular place.

With the passage of the Clean Water Act, water quality monitoring became institutionalized at the national level. Researchers and policy makers interested in comparing water quality from place to place led the initial attempts to establish a national water quality index. The U.S. EPA's monitoring program provides water quality data through a computerized data base which is available to most researchers. Unfortunately, spatial comparisons are often difficult to make. First, water quality standards are based on the functional use of the water; drinking water has more stringent quality standards than water used in irrigation or industry. Second, each water course is monitored on a drainage basin or individual stream level, thus comparability is limited at best. Not every stream or drainage basin has the same physical characteristics, flow regimes, vegetative cover, and assimilative capacity, all of which individually and collectively influence pollution levels. In addition, these geographic variations in physical characteristics may contribute to pollution levels in one area and exceed the recommended limits in another simply as a result of natural conditions. Third, many variables are needed to create a viable index. By oversimplifying the number, many important indicators are simply left out. Finally, it is virtually impossible to simplify the index into any coherent public warning system that is easily understood by the general public and policy makers.

Another attempt at a national pollution index is the national uniform air pollution index, commonly called the Pollution Standards Index (PSI). As with the water quality index, a national air index was first proposed in the early 1970s, yet did not become institutionalized until 1977. In an effort to standardize monitoring and reporting efforts,

TABLE 3   POLLUTION STANDARDS INDEX

PSI index value	Air quality level	Health effect label	Pollutant level				
			TSP (24-hour) μg/m^3	SO$_2$ (24-hour) μg/m^3	CO (8-hour) μg/m^3	O$_3$ (8-hour) μg/m^3	NO$_2$ (1-hour) μg/m^3
500	Significant harm		1000	2620	57.5	1200	3750
400	Emergency	Hazardous (PSI>300)	875	2100	46.0	1000	3000
300	Warning	———————	625	1600	34.0	800	2260
		Very unhealthful (PSI = 200 to 300)					
200	Alert	———————	375	800	17.0	400	1130
		Unhealthful (PSI = 100 to 200)					
100	NAAQS[a]	———————	260	365	10.0	240	(b)
		Moderate (PSI = 50 to 100)					
50	50% of NAAQS[a]	———————	75[c]	80[c]	5.0	120	(b)
		Good (PSI = 0 to 50)					
0			0	0	0	0	(b)

[a] NAAQS = National Ambient Air Quality Standard.
[b] There are no index values reported at concentrations below those specified by Alert criteria.
[c] Annual primary NAAQS.
Source: Council on Environmental Quality 1982:274.

the U.S. Environmental Protection Agency adopted a uniform air quality index in 1978. This index is a health-related measure and is based on the short-term national ambient air quality primary standards for five pollutants.

The PSI translates the concentrations of nitrogen oxides, sulfur dioxide, carbon monoxide, ozone, and total suspended particulates in metropolitan areas into a single value ranging from 0 to 500 (Table 3). When the levels for all five of the pollutants are below the primary standards the air is labeled "good" or "moderately polluted" (PSI values 0-99). When the ambient levels of any of the five pollutants exceed the primary standard, the PSI values range from 100-500 depending on the concentrations of pollutants in the air; the higher the concentrations, the higher the PSI value. An air quality alert is called when PSI values range from 100 to 200 ("unhealthful") and a warning is called when they reach the "very unhealthful" level (PSI 200-300). Along with each level, there are cautionary statements issued to the public on potential health impacts. Newspapers, radios, and television stations carry the readings, and statements such as "the air quality today is moderate" reflect these PSI readings.

Air and water quality are not the only areas of concern. Attempts to develop solid waste, hazardous waste, and noise indices were financed by the U.S. EPA, but none were ever fully developed and used, largely due to data unavailability and lack of uniformity. In addition to pollution-oriented indicators of environmental quality, a number of natural environment evaluations attempted to quantify scenic riverscapes, landscapes, and outdoor recreation (Leopold 1949; McHarg 1971; Bisselle *et al.* 1972;

Zube *et al.* 1975; Dearden 1980). The use of environmental impact statements (EIS), which define current environmental conditions and impacts of disturbance, is another limited attempt at a composite environmental quality assessment. Unfortunately, these impact statements cover extremely small areas, have widely-varying data bases, methodologies, and analytical techniques (Rau and Wooten 1980). The use of EISs to compare environmental quality from place to place is, therefore, inappropriate.

The enormity and complexity of the task severely limits the development of a comprehensive environmental quality index. Serious methodological problems arise in relating hundreds of factors in dozens of categories to one another. It is like comparing apples and oranges and the relative importance of each to a fruit salad. Aside from comparability, many pollution indicators are monitored on an infrequent basis and do not provide complete areal coverage.

One of the most widely-known composite environmental quality indices is published yearly by the National Wildlife Federation (NWF). NWF uses six natural resource categories — air, water, soil, forests, wildlife, minerals, and living space — as indicators. A quality rating ranging from 0 (bad) to 100 (good) is assigned to each subject. Data provided by government monitoring programs are analyzed by the NWF staff and consultants to determine overall quality in each subject area. Weights are then assigned to each subject area proportional to its relative importance in sustaining life. For example, air and water are more important for human existence than minerals and hence have a greater weight. The ratings on quality are then multiplied by the relative importance to determine the overall total of environmental quality (EQ) points (Table 4). These ratings are then compared to some ideal level of environmental quality or to previous years to assess positive or negative changes in environmental quality. Figure 2 illustrates the generalized environmental trends described by this index since 1970. According to the NWF EQ index, overall environmental quality declined in the U.S. from 1970-83 although air and forest resources have improved.

The NWF's environmental quality index essentially provides rudimentary information about trends in environmental quality. It lacks the scientific rigor of many of the indices described thus far. The quality ratings, based on the 'informed' opinions of staff and consultants, are more subjective than objective and may reflect biases of those providing the analysis. The value of the NWF index is its informed judgment about the general condition of the environment.

TABLE 4   NATIONAL WILDLIFE FEDERATION'S
ENVIRONMENTAL QUALITY INDEX

Category	Relative Importance	X	Rating	= EQ Points
Soil	30		78	23.40
Air	20		34	6.80
Water	20		40	8.00
Living Space	12.5		58	7.25
Minerals	7.5		48	3.60
Wildlife	5		53	2.65
Timber	5		76	3.80
	100.0			55.50

Source: Kimball 1972.

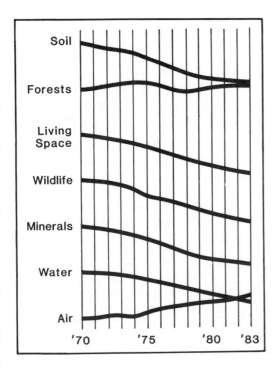

FIGURE 2   NATIONAL WILDLIFE FEDERATION'S EN-
VIRONMENTAL QUALITY INDEX. Trends in
environmental quality from 1970-1983 (see
text for description). Redrafted from *National
Wildlife* (February-March), 1983.

While we have stressed the near impossibility of constructing a comprehensive environmental quality index, there are many sources of environmental data which can be used to create separate indices of water, air, or other types of environmental quality. Data on virtually every aspect of the environment are available from either federal, state, or local agencies. The Council on Environmental Quality publishes an annual report on the quality of the environment where trends are noted. In addition, the U.S. EPA regularly monitors air and water quality as well as toxic substances. The U.S. Geological Survey also provides data on stream characteristics. Environmental protection agencies at the state and local level also provide a good source of environmental quality data. The data sources are too numerous to list or describe individually here, but suffice to say that most data are available in government publications.

The scale of environmental data is normally restricted spatially and is specific to the mission of the regulatory agency or monitoring program. Ambient air quality data, for example, are gathered by region or metropolitan area. Similarly, emissions data are monitored at the regional, state, or metropolitan level. As a consequence of federal pollution-control legislation, data bases are quite easy to compile. The only severe limitation is the scale of the analysis which hinders comparisons among different types of environmental indicators.

Perceptual Indicators

Perceptual indicators are most often used to assess the psychological well-being of individuals or people's evaluations of places or living conditions there. Most of the research on perceived environmental quality indices (PEQIs) falls somewhere in the continuum between group-centered studies at one end and place-centered studies at the other. The group-centered PEQIs focus on individual appraisals of well-being and seek to rank quality of life with respect to the perceptions of a particular group. The primary concern is comparing these quality assessments among groups of individuals. Are women more satisfied than men in terms of their overall quality of life? How does a person or group feel about those aspects of their residential or social environment? Place-centered PEQIs form an alternative approach. They are used to compare perceived quality of life among distinct places and are based on summary statistics of an aggregate of individual assessments of particular places. For example, would most Americans rank New York or Los Angeles as the better place to live? In all cases, however, PEQIs are based upon individual evaluations of the physical and built environment, including living conditions, problems, and so forth.

The stimulus for establishing psychological well-being as part of the overall assessment of quality of life came from the social indicators movement. In the early 1970s there was a small but influential group of social psychologists and sociologists who were increasingly dissatisfied with objective measures of social well-being. They felt that social well-being also meant feelings and satisfaction with one's quality of life. Campbell and Converse (1972) argued for the inclusion of perceptual indicators as necessary components to the understanding of the quality of American life. In their work they described the significant dimensions of psychological well-being such as personal happiness, life satisfaction, stress, and anxiety. They then suggested the types of measurement techniques that could be used to quantify these components, such as semantic differential scales. While Campbell and Converse's monograph was largely methodological and more conceptual in its orientation, it did provide the initial framework for more sophisticated and empirically-based studies.

In 1976, two major empirical studies of the perceived quality of life in America were published. Carrying on from their original methodological base, Campbell and Converse were joined by Rodgers (1976) in documenting the subjective quality of life in the U.S. They sampled 2,100 individuals and asked them to express their aspirations, evaluations, expectations, and overall feelings about their life. They found that subjective quality of life was a function of the satisfaction of basic needs. These needs were labeled "life domains" and included marriage, family life, health, life in the U.S., housing, standard of living and so on. Those life domains in which respondents claimed greater satisfaction included marriage and family, while dissatisfaction occurred in standard of living, housing, and city/county domains. One of the primary theses in the study concerned the relationship between objective measures of well-being and perceptual evaluations of it. Not surprisingly, Campbell, Converse, and Rodgers concluded that there is some divergence between objective indicators and subjective evaluations. They argued for the need for both subjective and objective indicators in determining overall quality of life.

Andrews and Withey (1976) published a similar study assessing social indicators of well-being. They also used a questionnaire survey which was administered to over 5,000 people nationwide. The primary purpose of the study was to develop actual measures of well-being and appropriate scaling techniques, thus enabling researchers to compare responses between individuals. Andrews and Withey also developed salient life domains or components of social well-being. These components included self, family, other people, economics, job, house costs, local area, larger society, and so on, and were measured on a "Delighted — Terrible" scale. Their results were similar to Campbell, Converse and Rodgers: respondents were most delighted with aspects of their private world (children, marriage, friends, self) and were least delighted with publicly-shared concerns such as the cost of goods and services, taxes, government. In addition, Andrews and Withey found very little variation in responses among racial groups, or between men and women. They did find significant variations among different socio-economic groups with respect to housing, economic, and local area components.

Subjective evaluations of social conditions and social well-being are not the only use of perceptual indicators. In 1976, Craik and Zube examined the research needs and methodological issues involved in evaluating physical environmental quality. Their primary interest was the link between an individual's perception of environment and the objective measures of environmental quality. For example, Weinstein (1976) reviewed research on how noise was perceived, the relationship between actual and perceived noise levels, and how this might be used in policy (setting community noise standards). Most of the early studies were exploratory in nature and examined general questions relating to perception of the environment, how individuals form environmental images, how they express these environmental images, and how they evaluate various aspects of the physical environment. There were a few empirical studies which attempted to link perceptual indicators to more objective measures of environmental quality (Jacoby 1972; Caris 1978; Cutter 1981).

There is considerable research in the fields of environmental psychology and behavioral geography on individual evaluations of both physical and built environments. Recent summaries of this type of interdisciplinary work can be found in Saarinen (1976), Downs and Stea (1977), and Saarinen and colleagues (1984), with periodic updates in *Progress in Human Geography* (Saarinen and Sell 1980, 1981). In addition, the geographical community has a long-standing interest in images of place, regions, and how they influence landscape perception, modification, and change (Tuan 1974; Lowenthal and Bowden 1976; Relph 1976; Sell et al. 1984). As a result of this interest, geographers contribute to both group-centered and place-centered PEQIs.

The focal point for place-centered PEQIs is the subjective assessment of a specific geographic locale by a group of people. Everyone has images about real and imaginary places. These images influence, to some degree, our subjective evaluations of place and, ultimately, quality of life. There are a number of factors which influence our individual images of place. These include individual factors such as personality, learning, knowledge, and experience (Canter 1983). These internal factors are influenced by a number of exogenous factors such as culture, mass communication, and the use of symbols to represent certain places. Yet, how are these images communicated to others?

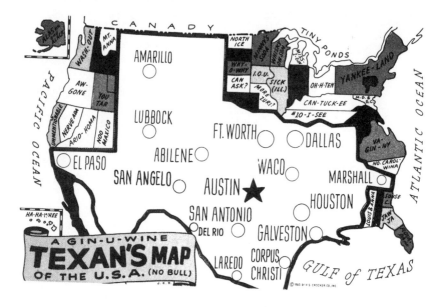

FIGURE 3   GEOGRAPHIC CHAUVINISM AND PLACE IMAGES. A
Texan's view of the U.S.A. (Postcard courtesy of Walcott
and Sons, San Angelo, Texas).

Geographic stereotyping is one method. Certain images of place are fostered by regional identification or vernacular regions which are both culturally and socially defined (Boney 1976; Zelinsky 1980; Garreau 1981). "The South" may not be one specific place that can be defined in strict geographic terms, yet it connotes a myriad of images about a place and the quality of life found there. The "East" may also be hard to delimit on a map, yet we can describe certain attributes of the place that might influence the quality of life — fast-paced, cold, dirty, urban, uptight.

Another way of communicating images of place is through geographic chauvinism. You may have seen the types of maps, posters, and postcards which represent "A New Yorker's View of the World" or the "Texan's View of the United States" (Figure 3). Almost every city now has these views which clearly typify their topophilia. Another method of communicating images of place is through the use of map postcards (Renwick and Cutter 1983). Postcard maps send simplified, yet sometimes erroneous, information about place and are often used to bolster a state's own image of itself as a tourist attraction.

Finally, the last way of communicating images of places to others is through the use of cognitive maps, which present a person's own representation of some portion of the real world (Downs and Stea 1977). An example is the sketch map you draw for directions to your house or the map of your college or university where you highlight places to see. Not only do our images of place and consequent evaluations of them have an important impact on our daily lives including where we choose to live, work, and play, but they also influence our comparisons and ratings of different places. As Downs and Stea remarked (1977:24):

*All of these variations in perspective emphasize that the world is what we
make it, that the world as we believe it to be depends upon our sensory
capacities, our age, our experience, and our attitudes and biases.*

Data on perceptual indicators are exclusively derived from surveys where individuals respond to a series of questions concerning their feelings, likes, dislikes,
aspirations, and attitudes about aspects of their life, including the attributes of the built
and physical environment. These individual responses are then summed to provide an
overall measure of individual well-being or societal feelings on these topics.

Subjective data on individual well-being are now routinely incorporated into a
number of governmental data bases. Both the 1977 and 1980 *Social Indicators*
volumes, for example, include sections on perceptual indicators of quality of life (Table
5). Neither, however, disaggregate the data to permit comparisons of perceived
well-being in different places.

In 1978, the U.S. Department of Housing and Urban Development (HUD) published a comprehensive survey on the quality of community life in the U.S. The survey
was commissioned in response to two needs:

*President Carter's pledge to formulate a comprehensive urban policy,
and his determination that citizens everywhere have input into the
deliberations that would produce such a policy (U.S. HUD 1978:3).*

The HUD survey conducted over 7,000 interviews with individuals from all over the
country representing various social, ethnic, income, and racial characteristics of the
population. In addition, the sample population was geographically stratified by city,
suburb, and rural locations. The resulting monograph had seven sections on various
aspects of quality of life ranging from community problems, to the performance of
government to residential and housing preferences (Table 5). To date, this is the most
massive undertaking by the federal government to assess the subjective dimensions of
quality of life in this country. Unfortunately, the spatial comparisons between places are
limited to city, suburb, and rural areas (the actual location of respondents) and
self-reported locations listed as large, medium, and small cities, medium and small
cities in suburbs, and rural areas.

Subjective evaluations of neighborhoods and residential environments are also
conducted by the U.S. Bureau of the Census. The Annual Housing Survey, for
example, not only measures the condition of the housing stock in sample cities, but also
solicits subjective assessments of quality. The survey, consisting of a national sample
of 70,000 housing units (about 5,000-15,000 units in sixty sample metroplitan areas),
provides information on existing housing stock, household composition, and housing
costs for the HUD (Dahmann 1981). The subjective indicators include overall satisfaction with the neighborhood; evaluations of the quality of local public services such as
shopping, public transportation, and schools; perceptions of the presence of specific
conditions in the neighborhood such as traffic and street crime; and, finally, evaluations
of the nuisance level of conditions they report (Dahmann 1981). The evaluations are
person-centered and have very little spatial information other than "neighborhood."
These measures are quite useful in determining residents' dissatisfaction with their
neighborhood and residential environment and their likelihood of moving, but tell us
little about evaluations of different places, other than on a generalized level such as
metro area, central city, and rural countryside.

TABLE 5   SELECTED SOURCES OF SUBJECTIVE DATA ON SOCIAL AND ENVIRONMENTAL QUALITY

Source	Number of Variables	Content Areas	Areal Coverage	Number of Places
U.S. Bureau	3	population	U.S.	1
of the Census	4	family life		
1977, 1980	4	evaluation of housing		
	1	social issues		
	3	health status		
	5	public safety		
	2	education		
	4	job satisfaction		
	3	consumer problems/ personal finance		
	4	leisure time activities		
	6	social class & anomie		
	39	Total		
U.S. HUD	54	quality of urban life	U.S.	3, 6[a]
1978	14	community problems	(4 regions)	
	11	social neighborhood		
	23	use of cities (work/play)		
	37	government performance		
	23	housing preference		
	28	community trends		
	190	Total		

[a]Three actual locations are given (central city, suburb, rural) which correspond to the place of residence of the respondent. Six additional locations are provided and are based on self-reported locations (large, medium, small city, medium city in suburbs, small city in suburbs, rural).

Mapping and Areal Differentiation

Many methods and models are used to measure the components of quality of life, yet few of them take the next step, comparing different areas or places. The use of quality of life components as a comparative spatial measure dates back to the 1920s when social indicators were first used to compare and distinguish social areas within cities.

Urban Social Areas

The urban ecological model was developed in the 1920s by urban sociologists at the University of Chicago to help explain the pattern and process of social differentiation within that city. These ecological studies provided detailed descriptions of areas in the city defined on the basis of the social, economic, and demographic characteristics of the area's inhabitants. These micro-studies of place formed the basis for community and neighborhood identification (Wirth 1928; Zorbaugh 1929; Suttles

1968, 1972; Hunter 1974). Areas of high status could be mapped and included those census tracts with individuals in professional occupations who were highly educated, had high incomes and who were mostly white (the Gold Coast in Zorbaugh's terminology — Figure 4A). Low status communities on the other hand were those areas (census tracts) where the residents had little education, were unemployed and whose income was below the poverty level (the "slum" according to Zorbaugh, the "ghetto" according to Wirth).

This approach to mapping social areas was used for some time because of its simplicity. Data were collected at the census tract level, which enabled researchers to replicate the study in different cities and then compare social areas of one city to another.

In the 1950s there was a new model of social differentiation proposed by Shevky and Bell (1955). Termed "social area analysis," this model was a reaction to the Chicago School and the urban ecological model which relied on socioeconomic status measures to differentiate areas. Shevky and Bell classified homogeneous groups of people with respect to 3 dimensions: social rank (economic status); family status (urbanization); and segregation (ethnic status). Social rank was measured by years of schooling, employment, value of home, persons per room, and so forth. Family status was measured by age and sex, owner or renter, and house structure. Segregation was measured by race and nativity, country of birth, age. These measures were placed into an additive index and were initially used to differentiate social areas within San Francisco (Figure 4B). Again there were criticisms of the rationale for the selection of variables as well as the use of census tract data to describe all individuals who lived there. In addition, opponents felt there was little attempt at understanding the process of residential or social patterning in a city.

The "factorial ecology" model was developed in the 1960s by urban geographers who were dissatisfied with the limitations of social area analysis (Murdie 1969; Rees 1970, 1971; Berry and Horton 1970; Berry and Kasarda 1977). Factor analysis, a statistical technique, was applied to the examination of social areas in order to identify the dimensions of social space or social indicators without any *a priori* assumptions about those dimensions. It was an attempt to statistically validate the implicit hypotheses in social area analysis. On the practical side, the use of the technique and the computer enabled researchers to increase the number of variables that were included in the analysis. Factorial ecology studies were done throughout the world in order to differentiate social groups and social space within cities (Figure 4C). The results of all of these studies suggest that social areas are distinguished by three dimensions; socioeconomic status, family or life cycle, and ethnicity or segregation. Some of these parallel the earlier notions of the Chicago School and Shevky and Bell.

During the 1970s, the term, "territorial social indicator," came into vogue, particularly in Britain (Knox 1975, 1978; Pacione 1982). Area profiles of places at a variety of scales were compiled using descriptive social statistics. These descriptive vignettes highlighted the spatial or territorial variations found in social conditions. This type of descriptive analysis of social conditions by place was used to guide area-based social and public policies in Britain and has become part of the set of analytical techniques used by urban and regional planners (Knox 1978). Territorial social indicators were not, however, the panacea that everyone had hoped. As Knox commented (1978:82):

It is worth stressing that area profiles and territorial indicators can be only

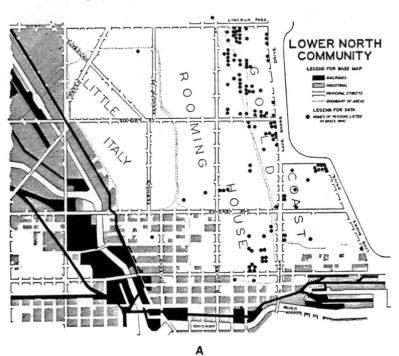

A

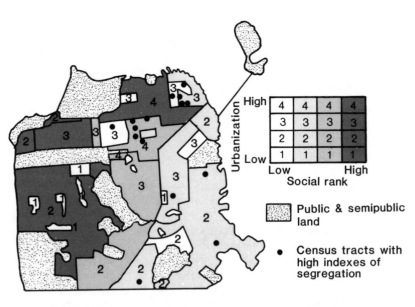

B

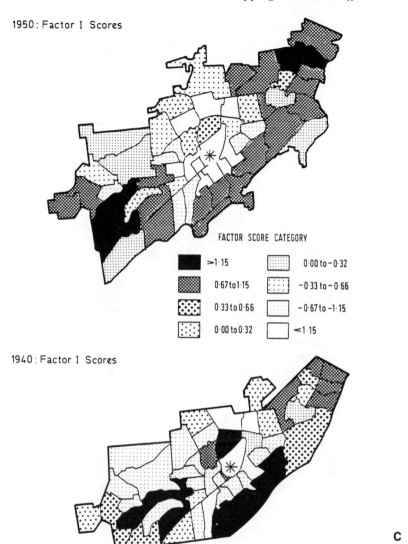

1950: Factor I Scores

FACTOR SCORE CATEGORY

■	>1·15	▦	0·00 to −0·32
▨	0·67 to 1·15	⠿	−0·33 to −0·66
⠿	0·33 to 0·66	⠿	−0·67 to −1·15
⠿	0·00 to 0·32	☐	<1·15

1940: Factor I Scores

C

FIGURE 4 URBAN SOCIAL AREAS. **A,** Zorbaugh's (1929) Gold Coast (reprinted with the permission of The University of Chicago Press); **B,** Shevky and Bell's (1955) social area analysis of San Francisco (redrafted with the permission of the publisher from *Social Area Analysis* by E. Shevky and W. Bell, Stanford University Press, 1955); **C,** Rees's (1979) factorial ecology of socioeconomic status in Birmingham, Alabama (reprinted with the permission of the Department of Geography, The University of Chicago).

as effective as their presentation allows. Unfortunately, the proliferation of technical jargon, the lack of comparative and contextual data. . ., and the lack of simple graphical and verbal summaries are all common attributes of recent reports and publications containing area profiles and territorial social indicators.

In the early 1970s, the Association of American Geographers sponsored a national atlas project describing and comparing America's largest urban places. This was an attempt to represent spatially the social conditions found in and among the nation's largest metropolitan regions. The resulting atlas (Abler and Adams 1976) provides a bevy of cartographic information on metropolitan areas as well as a narrative describing each. The first set of comparative maps described the historical development of each place. The second part of the atlas described the current (1970) conditions within the metropolitan area. Approximately 20 variables were mapped in five categories (the place, housing, people, socioeconomic characteristics, special topics) for each of the twenty areas. A narrative accompanied the maps for each metropolitan area. Finally, selected urban problems ranging from crowded housing, to premature births, to poverty, to households without cars were also mapped.

One of the indicators mapped, for example, is the proportion of household income which pays for rental housing. In 1970, 36.4% of the nation's renters paid more than one-fourth of their income for rent, but in some metropolitan areas it was much higher. The maps of rent in relation to income (Figure 5) not only illustrate inter-urban variations, but disparities within each metropolitan area as well. In Boston, more than 30 percent of the households pay one-fourth or more of their income for rent throughout the central city and most of the SMSA, while in Hartford, the percentages are greatest in and adjacent to the central city. This atlas then provided one of the first attempts to present systematically the spatial dimensions of social indicators in urban areas and comparisons between them.

Spatial Variations in Environmental Quality

While there are many studies on regional and national trends in environmental quality, very few have bothered to use environmental quality indices to differentiate places or compare environmental quality levels among places. The exception is the PSI which compares air quality in metropolitan regions nationwide. Trends in air quality can be monitored on a city scale and compared over time. The Council on Environmental Quality (CEQ) generally reports PSI data in this manner. For a spatial comparison, PSI values for metropolitan regions for a particular year are mapped (Figure 6). Graduated dots represent the number of days the PSI was in the hazardous or very hazardous categories. In this way, we have a visual representation of air quality in cities throughout the U.S. Using 1981 data, the worst air quality in the country is found in three areas — Los Angeles, San Bernadino-Riverside-Ontario, and New York. In each case, these regions had more than 100 days with PSI levels over 100. Some of the cleaner urban areas include Buffalo, Tucson, and Kansas City, each with less than ten days per year exceeding 100 on the PSI. This is one way to represent visually and compare air quality in metropolitan areas throughout the country.

One of the most comprehensive attempts at comparing environmental quality between places was Berry's *The Social Burdens of Environmental Pollution* (1977).

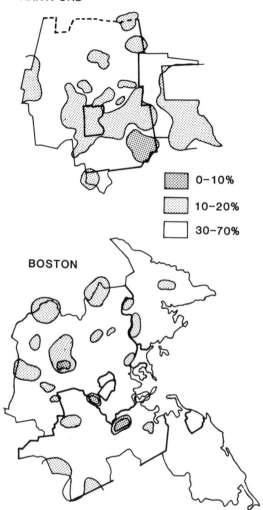

HARTFORD

0-10%

10-20%

30-70%

BOSTON

FIGURE 5   METROPOLITAN COMPARISONS OF INCOME TO RENT. Percentage of the household's income used for rent, 1970, in Hartford and Boston. Redrafted from *A Comparative Atlas of America's Great Cities: Twenty Metropolitan Regions,* 1976, published by the University of Minnesota Press for the Association of American Geographers. Used with the permision of the Association of American Geographers.

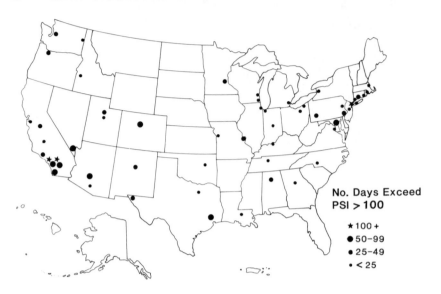

No. Days Exceed
PSI > 100

★ 100 +
● 50–99
● 25–49
• < 25

FIGURE 6   POLLUTION STANDARDS INDEX, 1981. Data from Council on Environmental Quality, 1982.

The primary research purpose was to ascertain whether different social groups were variously affected by pollution in metropolitan areas. Was it the inner city poor with limited mobility that were carrying the burden of pollution (being the most exposed to it) or was it the suburbanite?

Social and environmental quality data were collected for thirteen metropolitan regions (see Table 6). Maps showing the spatial pattern of the social or environmental indicator were then compiled, with an accompanying narrative describing the similarities and differences in quality. Twenty-four indicators of water quality and seven of air quality were mapped for each metropolitan region (Figures 7 and 8). Also represented on each map were the federal primary and secondary standards, so that the reader is instantly aware of whether or not a particular area met or exceeded the standards. Noise data were limited to four metropolitan areas and included only one indicator. Solid waste data were available for eight metropolitan regions and included five indicators.

Berry concluded that the minority poor are most afflicted with the greatest burdens of pollution. The affluent suburbanite, however, is also exposed to pollution. Suburban locations are increasingly confronted with new airports (increasing noise levels). They also rely on groundwater for drinking water which is depleted by overuse and becomes contaminated by improper solid waste disposal and the increased use of septic systems. Berry's volume represents one of the few attempts at comparing environmental quality among metropolitan regions and the impact of pollution levels on various social segments of these areas.

Place Images

One of the best examples of spatial comparisons using perceptual indicators is the preference ratings of college students on where they would like to live (Figure 9). The pioneering work on place preferences was first done in 1966 by geographer Peter Gould (Abler *et al.* 1971; Gould and White 1974). These subjective assessments of places (in this case, states) by students provided a visual representation of desirability. Most students preferred their own region over others, while certain areas such as the coastal West and Colorado consistently appear as preferred places regardless of where the students lived. Similarly, the least preferred places of all but the Alabama students were locations in the South. These maps do not provide information, however, on what aspects of these places were so desirable or undesirable, nor do they provide an absolute ranking of the best or worse places to live.

TABLE 6 TOPICAL AND AREAL COVERAGE IN BERRY (1977)

SMSA	Topics
Chicago	Water quality:
Baltimore	dissolved oxygen
Washington D.C.	biochemical oxygen demand
Providence	total coliform
St. Louis	fecal coliform
Rochester	pH
Cincinnati	total dissolved solids
Jacksonville	turbidity
Birmingham	methylene blue active substances
Oklahoma City	chloride
Denver	fluoride
Seattle	ammonia nitrogen
San Diego	nitrates and nitrites
	phosphorus
	sulfate
	cadmium
	copper
	iron
	lead, nickel, zinc
	Air quality:
	particulates
	sulfur dioxide
	nitrogen dioxide
	carbon monoxide
	hydrocarbons
	photochemical oxidants
	ozone
	Noise: ambient levels
	Solid waste: collection

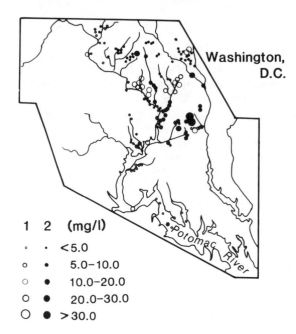

1   2   (mg/l)

○   ·   <5.0
○   ●   5.0–10.0
○   ●   10.0–20.0
○   ●   20.0–30.0
○   ●   >30.0

1 Frequent monitoring   2 Infrequent monitoring

FIGURE 7   URBAN VARIATIONS IN WATER QUALITY, BIOCHEMICAL OXY-GEN DEMAND. BOD measures the oxygen required to oxidize the organic material in surface waters and is used as an indicator of the presence of organic pollution from sewage or industrial wastes. Greater levels of organic pollution are found in Washington D.C. surface waters than in Rochester. (Redrafted from Berry 1977; used with the permission of the author).

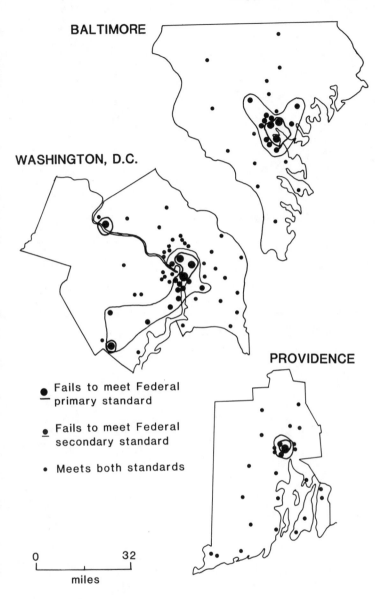

BALTIMORE

WASHINGTON, D.C.

PROVIDENCE

- Fails to meet Federal primary standard
- Fails to meet Federal secondary standard
- Meets both standards

0   32

miles

FIGURE 8 URBAN AIR QUALITY DIFFERENTIALS, PARTICULATES. Comparing air quality in Baltimore, Washington, D.C. and Providence shows higher particulate levels in the central cities than in the suburbs. (Reprinted from Berry 1977; used with the permission of the author).

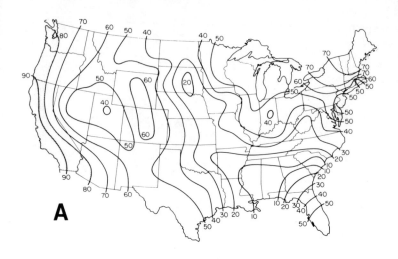

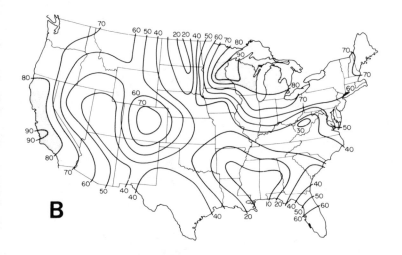

FIGURE 9   PLACE PREFERENCES. Residential desirability of states from college students in **A,** California; **B,** Minnesota; **C,** Alabama. Reprinted from Abler, Adams and Gould (1971) with the permission of the authors.

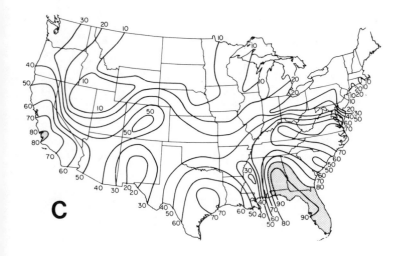

C

Summary

Determining the overall quality of life must include aspects of the physical, social, and perceptual worlds. Utilizing only one or two of these components provides an incomplete picture. It is difficult, if not impossible, however, to integrate perceptual information completely with more objective indicators of social and physical quality. Hence, there is a paucity of studies which attempt to do so. While the majority of research on quality of life indicators has focused on the methods and models used to measure them, there were attempts to use these indicators to define areas. Social, environmental, and perceptual indicators were used either singularly or in tandem to differentiate places and distinguish among attributes found in these places. In this regard, they provide a visual comparison of certain aspects of quality of life found in these areas, but they have to be used to rate or rank places. There are studies, whose sole purpose is to rank and compare places according to a number of dimensions to quantify the best and worst. While using questionable methodology, it is this type of research which produces the media interest in quality of life.

3

Ranking and Rating Places

Ratings are totally subjective. Some people like to visit museums, others do not; some like snow, others prefer tropics. We should resist the temptation to rank cities. Let people select their own favorite places.
— Letter to the Editor, *Time Magazine,* April 8, 1985

While some of the studies discussed in the previous chapter permit spatial comparisons of places along a number of dimensions (social quality, environmental quality), none of them ranked or evaluated these places from "best" to "worst." This chapter examines studies which use quality of life components to rank places. We begin at the largest spatial scale, international, and end with the smallest scale, intra-urban comparisons of neighborhoods.

There are a number of caveats to any discussion of a study that draws comparisons by ranking or rating key variables. Just as there are conflicting notions on how to measure quality of life, there are equally perplexing methodological problems in ranking or aggregating indicators into a summary measure. Statistically, indicators can, first, be aggregated and then compared using a variety of techniques ranging from simple linear summing to more complex multivariate and multi-dimensional techniques. One advantage of linear summing techniques is the ease of interpretation and analysis. Everyone understands what the mean, or average represents, and the higher or lower the average score, the better. Multivariate techniques require an understanding of the notion of independent and dependent variables, statistical concepts which may portray data more accurately, but are more difficult to understand. The primary advantage of these more sophisticated techniques is the large number and diversity of variables analyzed (see Ott 1978 for a discussion on the structure of indices and King 1969 and Wilson and Kirkby 1980 for a more general discussion of quantitative techniques in human geography). Keeping the caveat about variability among ranking methods in mind, how have people ranked places and which *are* the best and the worst?

Cross-National Comparisons: An International Perspective

For years, economists, sociologists, and international development agencies struggled with methods which would empirically define the well-being of nations in order to assess the effectiveness of international aid and development programs. The primary measure used for many years was per capita gross national product (GNP). There are however, many recognized problems with using only this indicator (Ram 1982). First, there are many difficulties with the computation of GNP, such as inadequate areal coverage (not all countries have appropriate data) or changes in the coverage over time, that restrict its use as a comparative measure. The second problem is the conversion of GNP from the unit of local currency to a common monetary unit such as the U.S. dollar. This automatically introduces a bias resulting from market exchange rates which vary dramatically over time. Also, a conventional GNP calculation neglects the contribution of subsistence agriculture and the underground economy of barter and black market trade to total productivity. Finally, there is some indication that a GNP calculation does not really measure social well-being, but simply income. This dissatisfaction with GNP as an indicator of well-being prompted a number of individuals and agencies to consider alternative measures of quality of life at the international level.

The Physical Quality of Life Index (PQLI)

The Physical Quality of Life Index (PQLI) was first developed by Morris (1979) for the Overseas Development Council and was used to supplement the per capita GNP measures in use at the time. The index consists of three indicators — life expectancy, infant mortality, and literacy. Each indicator is rated on a scale of 1 (worst) to 100 (best). The three indicators are then averaged (each has equal weight) to produce an overall score.

The Physical Quality of Life Index is used by international development agencies to classify countries according to the kinds of assistance they need or to measure the success of international aid efforts in improving quality of life. The PQLI index for 1981 covers approximately 164 countries. Of the top 10 countries, seven are in Europe while in the bottom 10, nine are African nations (Table 7). While the global average on the PQLI is 64, there is a wide disparity between developed countries (PQLI = 94) and developing countries (PQLI = 60). Highest quality of life is found in the countries of North America (PQLI = 96), Europe (PQLI = 90), and Oceania (PQLI = 87). In contrast, the lowest figures are found in Africa (PQLI = 37), Asia (PQLI = 62), and Latin America (PQLI = 76). It is interesting to note that in the latter three continents, only seven countries make it into the top 20% on the PQLI scale (Japan, Cuba, Barbados, Hong Kong, Israel, Soviet Union, Trinidad and Tobago) and only one of these makes it to the top 10 (Japan, rated second).

The spatial distribution of the PQLI suggests some correlation between a country's rating on the scale and its level of economic development (Figure 10). There is, however, a significant discrepancy in a number of regions where income or economic development does not ensure a good physical quality of life. Petroleum-rich Middle-Eastern nations have a high per capita GNP, yet rate very low on this index. Saudi Arabia, for example, had a per capita GNP of $3,529 yet its PQLI score is 35 which places it 132 out of 164. In contrast, the island country of Sri Lanka, with a very low per capita GNP ($179), had a PQLI of 80 and ranked 64th worldwide. While one may initially hypothesize about the link between GNP and quality of life, as measured by the PQLI, there are some divergent patterns.

One of the shortcomings of the PQLI is its reliance on only three indicators, all of which measure the health and educational levels of the country involved. The health indicators, for example, do not rely on measures of caloric intake, degree of medical services but use the very general measure of life expectancy to assess health. It does not even distinguish between male and female life expectancy. Despite these limitations, the PQLI does provide a crude comparative measure of the social or physical

TABLE 7   INTERNATIONAL RANKINGS

Physical Quality of Life Index			Index of Net Social Progress		
Rank	Country	PQLI Score	Rank	Country	INSP Score
1	Iceland	98	1	Norway	196
2	Japan	98	2	Denmark	194
3	Netherlands	98	3	Sweden	192
4	Sweden	98	4	Austria	189
5	Denmark	97	5	Netherlands	186
6	Norway	97	6	New Zealand	183
7	Switzerland	97	7	Australia	179
8	Australia	96	8	Finland	179
9	Canada	96	9	Ireland	176
10	France	96	10	Belgium	176
11	United States	96	11	Canada	174
12	Belgium	95	12	West Germany	170
13	Finland	95	13	Japan	165
14	Ireland	95	14	Switzerland	165
15	Italy	95	15	Poland	163
16	New Zealand	95	16	France	161
17	United Kingdom	95	17	Hungary	161
18	Austria	94	18	Czechoslovakia	158
19	East Germany	94	19	Romania	158
20	West Germany	94	20	Costa Rica	152
145	Mozambique	29	88	Zaire	51
146	Cameroon	28	89	Central African Rep.	50
147	Laos	28	90	Somalia	49
148	Zaire	28	91	Togo	46
149	Senegal	25	92	Benin	45
150	Chad	24	93	Malawi	38
151	Bhutan	23	94	Niger	37
152	Central African Rep.	23	95	Ghana	36
153	Gabon	23	96	Zimbabwe	35
154	Guinea	23	97	Zambia	35
155	Mauritania	23	98	Mali	34
156	Yemen Arab Rep.	22	99	Mauritania	34
157	Afghanistan	21	100	Pakistan	32
158	Ethiopia	20	101	Nigeria	28
159	Niger	19	102	Burkina Faso	26
160	Burkina Faso	19	103	Tanzania	23
161	Mali	18	104	Burundi	20
162	Angola	17	105	Uganda	19
163	Gambia	17	106	Chad	11
164	Guinea-Bissau	15	107	Ethiopia	10

Sources: Kurian 1984 and Estes 1984.

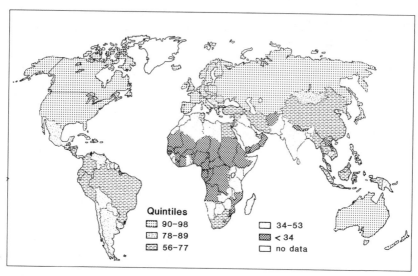

FIGURE 10   PHYSICAL QUALITY OF LIFE INDEX, 1981. See text for explanation. Data from Kurian (1984).

quality of life in nations and can be used to evaluate progress toward achieving social targets and goals, such as decreased illiteracy or increased life expectancy.

Index of Net Social Progress (INSP)

In response to criticisms of the PQLI, a multivariate index termed Index of Net Social Progress (INSP) was developed by Estes (1984), working as an academic consultant to the International Council on Social Welfare. This index, based on 11 subindices and using a total of 44 indicators, is more sophisticated than the PQLI (Table 8). Standardized scores are calculated for each variable and then summed for each subindex. The eleven unweighted subindices are multiplied by a constant factor of 10 (to increase the range of values) and then placed in an additive model to derive the overall score. Because of the restricted range of standardized scores, a constant factor of 100 is added to the computed overall score which is then rounded to the nearest whole number, creating an index of positive values. The overall INSP scores ranged from a low of 10 to a high of 196 for 1979-80.

Using data from 1979-80 for 107 countries, the INSP shows the preponderance of European nations in the top 10 (8 out of 10) and African nations in the bottom 10 (9 out of 10; Table 7). The only non-European countries in the top twenty percent are Canada, Costa Rica, Japan, Australia, and New Zealand. The countries in the lowest quintile are in Africa except for Pakistan. As seen in Figure 11, there is a definite developed versus developing nation dichotomy in the spatial pattern of the rankings. Another interesting pattern is the rankings of some of the world's foremost powers. The U.S., for example, ranks 23 with a score of 146, while the United Kingdom ranks 28 (INSP = 139) and the Soviet Union 45 (INSP = 113).

Because this index de-emphasizes economic indicators (only 4 out of 44 variables measure wealth) and includes criteria — such as political participation, political stability, and women's rights in addition to the traditional education and health indicators — the U.S., in particular, drops out of the top 10. This is largely a result of differences between minority groups and whites which are reflected in the health status index and the cultural diversity index. In addition, political protests, terrorist attacks, and deaths from domestic violence during the mid to late 1970s gave the U.S. a lower rating on the political stability subindex. Finally, the number of natural disasters affecting the U.S. and the damage from them lowered the U.S. score on the geographic subindex.

Rating Nations

While both the PQLI and INSP have limitations, their list of top 10 nations and bottom 10 nations were similar. Both indices concluded that the highest quality of life was found in the Scandinavian countries of Sweden, Denmark, and Norway, plus the Netherlands and Australia. Similarly, the lowest quality of life was in the African countries of Ethiopia, Mali, Mauritania, and Burkina Faso where the current physical quality of life was low and progress toward improving social conditions was equally slow. Both indices ignored the other two components of quality of life — environmental and perceptual — and thus only represent the objective social conditions that were found in each nation. The INSP does make a furtive attempt to use an environmental index through its geographic subindex. This measures arable land and the number of natural hazards and human vulnerability to them. While it may be a start, this geographic subindex by no means provides a comprehensive picture of the environmental component of a nation's quality of life.

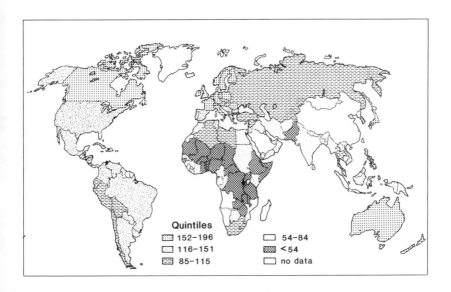

Quintiles
- 152–196
- 116–151
- 85–115
- 54–84
- <54
- no data

FIGURE 11   INDEX OF NET SOCIAL PROGRESS, 1980. See text for explanation. Data from Estes (1984).

TABLE 8   INDICATORS USED IN THE INDEX OF NET SOCIAL PROGRESS
(INSP)

I **Education Subindex**
   school enrollment ratio, first level (+)
   pupil teacher ratio, first level (-)
   percent adult illiteracy (-)
   percent GNP in education (+)

II **Health status subindex**
   infant mortality rate (-)
   population in thousands per physician (-)
   male life expectancy at 1 year (+)

III **Women status subindex**
   percent age eligible girls attending first level schools (+)
   percent girls in primary schools (+)
   percent adult female illiteracy (-)
   years since women suffrage (+)
   years since women suffrage equal to men (+)

IV **Defense effort subindex**
   percent GNP in defense spending (-)

V **Economic subindex**
   economic growth rate (+)
   per capita estimated income (US$) (+)
   average annual rate of inflation (-)
   per capita food production index (1970 = 100) (+)

VI **Demography subindex**
   total population (1000) (-)
   crude birth rate per 1000 population (-)
   crude death rate per 1000 population (-)
   rate of population increase (-)
   percent of population under 15 years (-)

VII **Geography subindex**
   percent arable land mass (+)
   number major natural disaster impacts (-)
   lives lost in natural disasters per million pop. (-)

VIII **Political stability subindex**
   number of political protest demonstrations (-)
   number of political riots (-)
   number of political strikes (-)
   number of armed attacks (-)
   number of deaths from domestic violence (-)

IX **Political participation subindex**
   years since independence (+)
   years since most recent constitution (+)
   presence of functioning parliamentary system (+)
   presence of functioning political party system (+)
   degree of influence of military (-)
   number of popular elections held (+)

TABLE 8   (continued)

X	**Cultural diversity subindex**
	largest percentage sharing same mother tongue (+)
	largest percent sharing same basic religious beliefs (+)
	ethnic-linguistic fractionalization index (-)
XI	**Welfare effort subindex**
	years since first law — old age, invalidity, death (+)
	years since first law — sickness and maternity (+)
	years since first law — work injury (+)
	years since first law — unemployment (+)
	years since first law — family allowances (+)

Source: Estes 1984:23-24.
Note: + and - are directional indicators which refer to gains and losses and are either added to or subtracted from the subindex total.

Comparing States in the U.S.

There have been a variety of studies conducted on inter-state comparisons of places ranging from the very general (Conway and Liston 1981) to more sophisticated analyses (Smith 1973). These studies reflect a long-standing American fascination with regional and state identity as reflected in the image of a state by others, as well as competition among states for everything ranging from tourists, to new industry, to professional sports teams. Most of these inter-state studies focus on social indicators only, although occasionally perceptual indicators are used. There are few, if any, inter-state studies on quality of life using environmental indicators either singularly or in conjunction with social and perceptual measures.

H.L. Mencken's "Worst American State"

In the September 1931 issue of *The American Mercury,* Charles Angoff and H.L. Mencken provided a unique study of the social quality of life in the U.S. They wanted to determine the states that were the "best" and "worst" places to live, and whether or not regional stereotypes still persisted. As Angoff and Mencken commented (1931:2):

> *Not many natives of New England would admit that Iowa is as civilized as Massachusetts, and not many Southerners would admit that either the North or the West is as civilized as the South.*

The authors used 63 variables to create four general social indices — wealth, education, health, and public order. Variables ranged from income, educational attainment and savings, to magazine circulations to numbers of lynchings. The 48 states and the District of Columbia were ranked from first (best) to forty-ninth (worst) for each variable. Where two or more states tied, they were assigned the same rank and the next ranking was lowered accordingly to maintain 49 places. Twenty-six variables were used to create the wealth index, 24 for education, 11 for health, and 2 for public order.

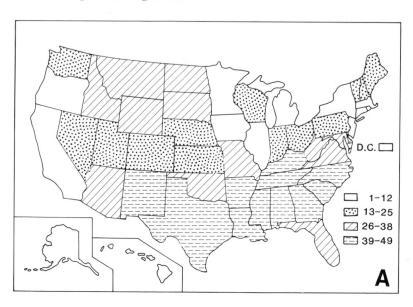

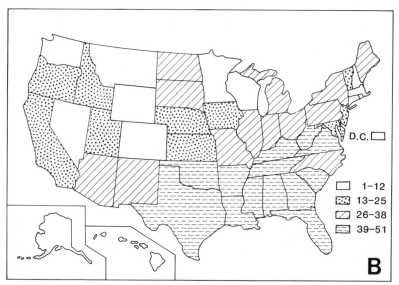

FIGURE 12   INTER-STATE RANKINGS, 1930 AND 1980. **A,** 1930
quintiles based on Angoff and Mencken (1931); **B,** 1980
quintiles based on Cutter *et al.* (1985); **C,** Changes over
the 50 year period. Western states register largest gain
and snowbelt northern states the largest loss in standing.

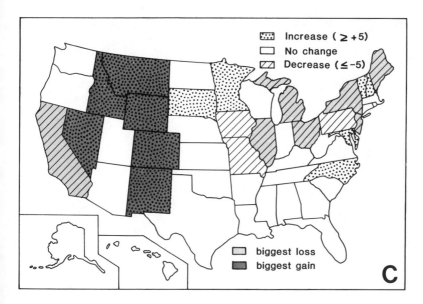

The state rankings for each variable were then arithmetically averaged for each subindex. These subindices were, in turn, ranked, summed, and averaged to provide the overall score which determined the "best" and "worst" state.

The top and bottom ten states in each of the four indices are listed in Table 9. Northeastern and western states consistently appear in the top ten in all four categories, while southern states consistently appear in the bottom ten ranks for each subindex. Angoff and Mencken concluded that the "best" state in the U.S. was Massachusetts, followed by Connecticut, New York, New Jersey, and California and that the "worst" states were Mississippi, Alabama, South Carolina, Georgia, and Arkansas (Table 9). Spatially, the pattern is quite distinct (Figure 12A). The south is uniformly ranked at the bottom, while the states in the northeast, midwest, and coastal west ranked at the top.

The original Angoff and Mencken study hypothesized regional variations in social progress and found them. There have been substantial advances in health, wealth and education throughout the country and an increasing parity among regions since 1931. If Angoff and Mencken's conducted their analysis today, would the results be the same?

Angoff and Mencken's methodology was replicated with 1980 data using the same system of arithmetic averaging, and as nearly as possible, the same variables (Cutter et al. 1985). Variables were modified, dropped or added because of data availability and/or obsolescence. For example, one of Angoff and Mencken's wealth variables was the number of telephones per state. Since most households now have telephones, the number of cable television subscribers per state was substituted. In another instance, Angoff and Mencken used two variables in the education subindex which are now obsolete: urban illiteracy and decrease in illiteracy 1890-1930. Since all children are required to attend grade school, where they are presumably taught to read and write, illiteracy statistics are no longer reported.

TABLE 9   ANGOFF AND MENCKEN INTER-STATE RANKINGS, 1931 AND 1980

Rank	Wealth 1931	Wealth 1980	Education 1931	Education 1980	Health 1931	Health 1980	Public Order 1931	Public Order 1980	Overall 1931	Overall 1980
Top Ten										
1	CA	WY	MA	MA	MN	HA	ME	DE,ME, NC,SD, VT,WY	MA	CT
2	CT	CT	DC	NY	MA	RI	VT		CT	MN
3	NY	MN	CA	CA	NY	CT	NH		NY	CO
4	DC	IA	CT	DC	CT	VT	MA		NJ	MT
5	MA	DE	MI	CO	IA	MA	RI		CA	MA
6	IL	CO	CO	WA	OR	CO	CT		DC	DE
7	NV	IL,CA	UT	MI	NJ	NV	NJ	RI,NH, MT,ID	MN	RI
8	NJ		IL		IL	MN	WI		IA	WY
9	RI	NV	NY	MN	NE	NH	MN		IL	NV
10	PA	KS	OR	PA	WA	OR	IA		OR	NH
Bottom Ten										
40	KY		KY		VA		KY		NM	
41	OK		TN		AZ		SC		KY	
42	TN	ME	NM	GA	LA	FL	AR	LA	LA	FL
43	NC	GA	NC	SD	GA	OK	TN	GA	NC	TX
44	LA	WV	LA	SC	TN	NC	AL	PA	TN	KY
45	GA	NC	SC	TN	NM	AR	NV	OH	AR	AR
46	AL	KY	GA	WV	NC	MS	GA	IL,MI	GA	TN
47	SC	AR	AR	AL	AL	SC	MS		SC	SC
48	AR	TN	AL	AR	MS	AL		FL	AL	MS
49	MS	AL	MS	LA	SC	TX		NY	MS	AL
50		MS		KY		LA		TX		GA
51		SC		MS		GA		CA		LA

Source: Angoff and Mencken 1931; Cutter et al. 1985.

As one might expect, there are both similarities and differences between the Angoff and Mencken update and the original study. For example, there are many changes in the top ten states, but very few in the bottom ten. The South as a region is still consistently low in all of the individual indicators ranging from wealth to public order (Table 9). The only variation of this pattern occurs in the public order subindex where the northern industrialized states with large urban populations, plus California, consistently rate at the bottom of the scale. The top ten states in each of the subindices are quite disparate and range from coast to coast. Spatially, the top ten states are concentrated in New England and in the Mountain West (Figure 12B) which is somewhat different than the pattern in 1930 (Figure 12A). The bottom ten states are still located in the South, except the south as a region has been enlarged to include Florida and moved west to include Texas and Oklahoma.

There were quite a few surprises in the changes in the rankings over the fifty year period. The biggest regional net gain occurred in the Mountain West states which increased their ratings from the twenties and thirties to the top ten (Figure 12C). Montana, for example, was ranked number 31 in 1931 and jumped to 4th place in 1980 (Table 9). The largest net increase in place ratings occurred in Montana (+27),

Wyoming (+ 23), Delaware (+ 20), and Colorado (+ 20). In contrast, the biggest losers were the states in the industrialized East, Midwest, and California. Individually, New York fell dramatically from 3rd in 1931 to 30th in 1980. Likewise, Illinois dropped from 9th to 31st, while Ohio went from 17th to 33rd, California from 5th to 20th, and New Jersey from 4th to 19th.

The Angoff and Mencken study provided a popularized image of "best" and "worst" American states and gained a wide audience. The variables chosen reflect a certain bias (toward urbanism and wealth) and this helps to account for the low showing of rural and agricultural states in 1931. In the 1980 update, curiously enough, those states with large urban concentrations drop down in their rankings as a result of public disorder (murders, rapes) and wealth measures (number of people on public assistance). Angoff and Mencken did not choose any perceptual or environmental variables, and thus two-thirds of the components to quality of life are missing. Their work, and the 1980 update of it, does provide a glimpse of the social quality of life in American states and changes in the quality over 50 years.

Smith's Inter-State Analysis

As part of a longer monograph on the geography of social well-being in the United States, Smith (1973) examined territorial social indicators at a number of different scales. His inter-state analysis used 47 variables covering six general areas including income, wealth, and employment; environment; health; education; social disorganization; and alienation and political participation. Smith used the conceptual definitions of social well-being culled from the social science literature in his development of territorial social indicators. He readily admitted shortcomings in the analysis, largely as a function of data unavailability. For example, measures of the physical environment were omitted or restricted to only housing indicators. In addition, there were no perceptual indicators, as data are also unavailable. While recognizing the limitations of his research with respect to theoretical constructs, Smith nevertheless presented a respectable analysis of inter-state variations in social well-being.

Two different analyses of the data were used to determine state rankings — a standard-score, additive model and a principal components model. The additive model produced standard Z-scores on each of the six criteria which were then summed to produce a composite score. The top and bottom 12 states in each of the six criteria were then mapped (Smith 1973:87-88). The South was consistently in the bottom rankings on all the criteria except for social disorganization, where the West and Southwest were the lowest. The pattern for the top 12 states showed no clear trend although Connecticut and Massachusetts appeared in the top 12 in five categories and Utah, Washington, and Wisconsin in 4 of 5 categories. According to Smith's analysis using this additive model, Connecticut had the highest social well-being and Mississippi, the lowest (Table 10). The composite map showed some regional disparities (Smith 1973:87). The southern states are relegated to the bottom quartile, while the top twelve included states in New England, the Northeast, northern Midwest, West and Northwest.

The principal components analysis (a multivariate statistical technique) used the same 47 variables, and allowed the classification of subindices to occur statistically. Three factors were generated as the leading criteria (or components) which explained variations among states. These Smith labeled "general socio-economic well-being,

TABLE 10   SMITH'S INTER-STATE RANKINGS

State	Additive Model	Principal Components	State	Additive Model	Principal Components
Arizona	31	35	Nebraska	22	17
Arkansas	45	43	Nevada	32	30
California	16	22	New Hampshire	12	6
Colorado	11	15	New Jersey	7	20
Connecticut	1	3	New Mexico	35	36
Delaware	18	23	New York	9	28
Florida	38	39	North Carolina	43	42
Georgia	44	46	North Dakota	20	12
Idaho	14	8	Ohio	24	27
Illinois	23	29	Oklahoma	34	31
Indiana	30	26	Oregon	10	9
Iowa	4	2	Pennsylvania	21	24
Kansas	13	14	Rhode Island	17	18
Kentucky	40	37	South Carolina	47	47
Louisiana	42	44	South Dakota	27	19
Maine	29	21	Tennessee	41	41
Maryland	28	33	Texas	39	40
Massachusetts	2	11	Utah	3	1
Michigan	19	25	Vermont	25	12
Minnesota	5	5	Virginia	37	38
Mississippi	48	48	Washington	8	10
Missouri	33	32	West Virginia	36	34
Montana	26	16	Wisconsin	6	7

Data from Smith 1973.

social pathology, and mental health." By mapping the component scores, Smith was able to describe the spatial variations in social well-being. As was the case with the additive model, the southern states consistently appeared at the bottom of the general socio-economic well-being while the northeast and California appeared near the top. This pattern was somewhat reversed in examining the social pathology indicator. Here, the West (California and Nevada) and East (Maryland and New York) appeared on the bottom and the Plains and Rocky Mountain states were at the top. As Smith concluded (1973:103):

> To achieve a good overall or aggregate performance a state needs a prosperous agricultural economy, and urban areas without ghettos and barrios or which are predominantly middle-class dormitories. The combination of rural poverty and inner-city social problems ensures a state a position near the bottom of the scale.

Despite using different variables and statistical techniques, the findings of Angoff and Mencken, the 1980 update of it, and Smith's studies show remarkably similar results on a general regional basis. The South consistently appears at or near the bottom of the rankings on social well-being, while the Northeast appears at or near the top. It should be cautioned, however, that these studies only consider social indicators. With the inclusion of environmental and perceptual indicators, these rankings would surely change.

Regional Well-Being

Another less-rigorous study on regional quality of life provides a perceptual analysis of well-being. Using existing data from published research reports by the Institute for Social Research at the University of Michigan, Rubenstein (1982) re-evaluated survey results to look for regional variations in psychological well-being. She used 39 questions from these surveys which were then aggregated into six indices ranging from general outlook on life, stress, and personal competence, to positive and negative feelings and overall satisfaction. The regions chosen corresponded to Bureau of the Census reporting areas.

Rubenstein found a significant number of regional variations in well-being. The highest level of psychological well-being was found in the West South Central states (Oklahoma, Arkansas, Texas, Louisiana) followed by states in the West North Central region (Minnesota, North and South Dakota, Nebraska, Iowa, Kansas and Missouri; Table 11). The lowest psychological well-being was found in the East North Central and Mid-Atlantic regions. Rubenstein found that the psychologically healthiest individuals are not limited to the more affluent or sunbelt regions of the country, but those regions with a slower pace of life. In concluding, Rubenstein suggested that psychological well-being was a greater influence on the decision to migrate (along with employment

TABLE 11   PSYCHOLOGICAL WELL-BEING[a]

Region	States	Rank	Outlook on Life	Stress	Feelings Pos.	Feelings Neg.	Personal Competence	Life Satis.
West So. Central	OK,AR,TX, LA	1	2	1	3	7	3	2
West No. Central	MN,ND,SD, NE,IA,KS, MO	2	7	2	4	1	2	3
New Eng.	ME,NH,VT, MA,RI,CT	3	1	7	7	2	5	1
Mountain	MT,ID,WY NV,UT,CO, AZ,NM	4	6	6	1	4	1	6
Pacific	WA,OR,CA, HI,AK	5	5	3	2	8	4	8
South Atlantic	DE,MD,DC, VA,WV,NC, SC,GA,FL	6	4	4	5	5	7	5
East So. Central	KY,TN,MS, AL	7	3	5	9	3	9	4
East No. Central	MI,OH,IN IL,WI	8	9	8	6	9	6	7
Mid. Atl.	NY,NJ,PA	9	8	9	8	6	8	9

[a]Scores range from highest (1) to lowest (9). Data from Rubenstein 1982.

opportunities) than was previously recognized. She claimed it accounted for some of the large population outmigration from the psychologically depressed areas of the East North Central and Mid-Atlantic states to the more psychologically healthy regions in the South Central and Mountain states.

Rubenstein's work was a commendable effort in attempting to define regional variations in psychological well-being, but it provided no clear discussion of methodology, including how the subindices were constructed. How closely do the data represent the subindices Rubenstein used when they were obstensibly collected for other purposes? While there may, in fact, be regional variations in psychological well-being, implying that it is now a major influential factor in migration is somewhat speculative. This article represents an example of the popularized form of quality of life research that has questionable scientific validity, yet receives considerable media attention.

Urban Places: Cities and Metropolitan Areas

Interest in urban places, both by researchers and the general public, has stimulated quite an extensive number of empirical studies on various aspects of quality of life in these areas, ranging from crime indicators to general descriptions of prevailing social conditions. There are, however, relatively few empirical studies which focus on comparative measures of quality of life. In some instances, only one of the quality of life components — social indicators — was used, while in others, multiple components (social, environmental, perceptual or some combination thereof) were employed to compare quality of life among urban places.

Thorndike's Your City

In 1939 E.L. Thorndike published his compendium of recorded facts on American cities over 30,000 in population. The book examined the quality of life ("goodness of life," using Thorndike's terminology) in each of these cities and was one of the first comparative studies of quality of life at the inter-urban scale. It not only used the standard types of social indicators, but near the end of the monograph, Thorndike attempted to compare cities on the basis of perceptual indicators as well.

Thirty-seven variables were used in the analysis of 310 urban places over 30,000 in population. These variables included 5 health measures *(e.g.* infant death rates), 8 education measures *(e.g.* per capita public expenditures for schools), 2 recreation measures *(e.g.* per capita public park acreages), 8 economic and social measures *(e.g.* rarity of extreme proverty, per capita number of homes owned), 5 comfort indicators *(e.g.* per capita domestic installation of electricity), 3 literacy measures *(e.g.* per cent illiterates), and, finally, 6 measures of social conditions *(e.g.* per capita deaths from homicide).

Thorndike used a simple linear additive technique in determining rankings. The deviation from the score of the median city on each variable was calculated for all cities. The number of variables below the median were subtracted from the number of variables above the median for each city. These scores were then weighted by giving each city a bonus for excesses of plus over minus scores or a penalty for excesses of minus over plus scores. The result is the "general goodness of life" score for each city which Thorndike labeled "G-Scores."

These G-scores were then ranked in descending order with a high G-score indicating a good quality of life for that city. The top and bottom twenty-five cities are listed in Table 12. Pasadena was ranked as the number one city with the "best" quality of life and five cities were tied for the "worst" (Augusta and Columbus, GA; Meridian, MS; High Point, NC; and Charleston, SC). Spatially, the highest-ranked cities are concentrated in northern and southern California, and in the New York City metropolitan region (Figure 13). The lowest-ranked cities are found distributed throughout the South with the heaviest concentration in North Carolina. It is interesting to note that only one northern city, Lewiston, ME, ranks in the bottom 25 while no southern city ranks higher than 232 (or the lowest quartile of all cities).

Thorndike's original study was replicated for smaller cities with total population between 20,000-30,000 (Thorndike 1940). He found just as great a disparity in the goodness of life among these smaller cities as he did in the larger ones. The top-ranked smaller cities were Alhambra and Huntington Park, CA (tied for first place), while the lowest-ranked was Vicksburg, MS.

In addition to these ratings on social indicators, Thorndike also conducted a peripheral analysis of the perceptual ratings of the larger cities to show that the

TABLE 12   THORNDIKE'S INTER-URBAN RANKINGS

Top	Twenty-five		Bottom	Twenty-five
1	Pasadena, CA		310	Augusta, GA
2	Montclair, NJ			Columbus, GA
	Cleveland Heights, OH			Meridian, MS
4	Berkeley, CA			High Point, NC
	Brookline, MA			Charleston, SC
6	Evanston, IL		305	Savannah, GA
	Oak Park, IL			Durham, NC
8	Glendale, CA		303	Montgomery, AL
	Santa Barbara, CA			Macon, GA
	White Plains, NY			Jackson, MS
11	Santa Monica, CA			Columbia, SC
12	Long Beach, CA			Laredo, TX
	Lakewood, CA		298	New Orleans, LA
14	Alameda, CA			Wilmington, NC
	Newton, MA			Winston-Salem, NC
	New Rochelle, NY		295	Mobile, AL
	East Cleveland, OH			Pensacola, FL
18	Oakland, CA			Shreveport, LA
	San Jose, CA			Chattanooga, TN
	East Orange, NY		291	Paducah, KY
21	Los Angeles, CA			Lewiston, ME
	Santa Ana, CA			Knoxville, TN
	Colorado Springs, CO		288	Little Rock, AR
	Mount Vernon, NY			Charlotte, NC
25	San Diego, CA			Raleigh, NC
	Springfield, MA			Memphis, TN

Data from Thorndike 1939.

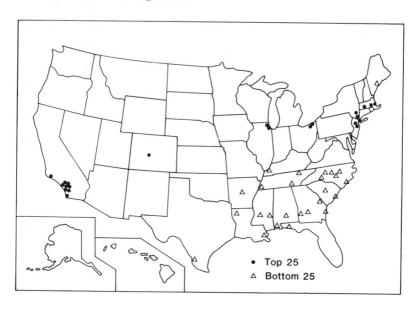

FIGURE 13   INTER-URBAN RANKINGS, 1939. Thorndike's (1939)
ratings of cities over 30,000.

reputation of a city is as important to its residents as the reputation of a family is to its
members (1939:143). He surveyed the opinions of 268 civic, business, education, and
religious leaders, asking their evaluations of the goodness of life in 117 cities. These
opinions were then compared to the more objective G-scores. Unfortunately, these
evaluations were not listed separately for each city, so we are unable to compare them
with the objective ratings for specific places. Thorndike did, however, draw several
conclusions about the relationship between objective measures of cities and subjective
evaluations of them. First, he explained that opinions regarding cities have little relation
to the objective conditions found there and that, very often, such opinions reflect one's
own quality of life, not the quality of life of others. Second, after correlating the opinions
with the G-scores he found that educators were closer to the objective ratings than
business leaders and clergy. Finally, and perhaps more aptly, Thorndike stated
(1939:150):

> The moral of this chapter as a whole is: Do not take anybody's opinion
> about your city; get the facts.

Liu's QOL in Metro Regions

The seminal work on inter-urban quality of life indicators was done by Liu (1975a).
The purpose of this study was to assess the quality of life and its variations among 243
metropolitan areas (then, SMSAs) in the U.S. This study, particularly its use of
multivariate and multidimensional techniques, provided the initial methodology for
subsequent QOL research. Five subindices were constructed (economic, political,

environmental, health and education, and social indicators) using 123 variables. Because of the inverse relationship between city size and quality of life noted in the literature (USEPA 1973b; Elgin *et al.* 1974), Liu disaggregated the SMSAs into three groups based on SMSA population (greater than 500,000; 200,000 to 500,000; less than 200,000) which he then labeled "large," "medium," and "small."

For each subindex, the data were transformed and their distributions normalized using Z-scores, thereby eliminating differences between the units of measurements (for example, comparing dollars with population). The standardized scores were then subjected to a factor analysis (a multivariate technique) to simplify the number of variables, and create a composite subindex. Each subindex was then placed into an

TABLE 13   LIU'S RANKINGS OF METROPOLITAN AREAS

	Top (Outstanding)			Bottom (Substandard)
	Large SMSAs			
1	Portland, OR	65		Jersey City, NJ
2	Sacramento, CA	64		Birmingham, AL
3	Seattle-Everett, WA	63		New Orleans, LA
4	San Jose, CA	62		San Antonio, TX
5	Minneapolis-St. Paul, MN	61		Jacksonville, FL
6	Rochester, NY	60		Greensboro-Winston-Salem-High Point, NC
7	Hartford, CT	59		Norfolk-Portsmouth, VA
8	Denver, CO	58		Memphis, TN
9	San Francisco-Oakland, CA	57		Philadelphia, PA
10	San Diego, CA	56		Tampa-St. Petersburg, FL
	Medium SMSAs			
1	Eugene, OR	83		Mobile, AL
2	Madison, WI	82		Charleston, SC
3	Appleton-Oshkosh, WI	81		Macon, GA
4	Santa Barbara, CA	80		Montgomery, AL
5	Stamford, CT	79		Columbus, GA
6	Des Moines, IA	78		Fayetteville, NC
7	Lansing, MI	77		Greenville, SC
8	Kalamazoo, MI	76		Columbia, SC
9	Ft. Wayne, IN	75		Huntington-Ashland, WV
10	Ann Arbor, MI	74		Augusta, GA
	Small SMSAs			
1	La Crosse, WI	95		Laredo, TX
2	Rochester, MN	94		Pine Bluff, AR
3	Lincoln, NE	93		McAllen-Pharr-Edinburg, TX
4	Topeka, KS	92		Ft. Smith, AR
5	Green Bay, WI	91		Lawton, OK
6	Ogden, UT	90		Brownsville-Harlingen-San Benito, TX
7	Norwalk, CT	89		Albany, GA
8	Sioux Falls, SD	88		Tuscaloosa, AL
9	Fargo-Moorhead, ND	87		Savannah, GA
10	Bristol, CT	86		Gadsden, AL

Data from Liu 1975a.

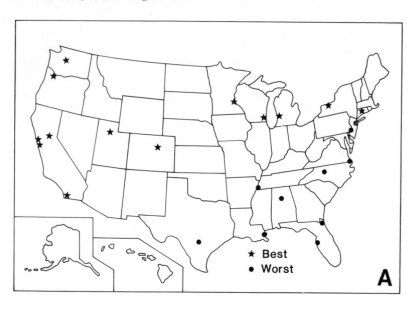

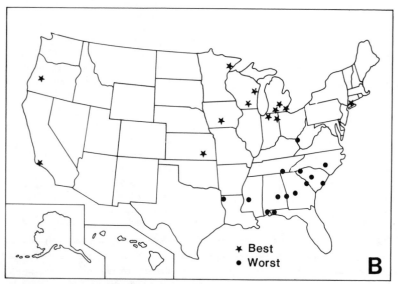

FIGURE 14 QUALITY OF LIFE IN METROPOLITAN AREAS, 1970. Liu's (1975) analysis of overall quality of life in **A,** Large SMSAs; **B,** Medium-sized SMSAs; and **C,** Small SMSAs. See text for description.

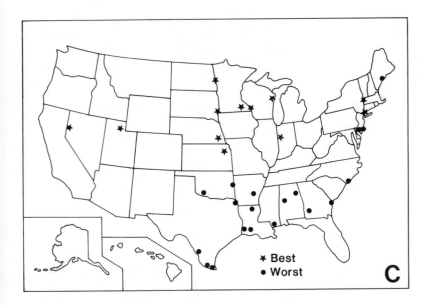

additive model to derive the overall score. While this description of his methodology is an oversimplification, it does provide some basic information on how Liu went about collapsing and analyzing a very large data base and ultimately deriving his score on each component and overall ratings (see Liu 1975a for a more detailed discussion of the methodology).

For the three metropolitan size categories, Liu provided a comprehensive assessment of each of the five components including an SMSA's overall score as well as letter rankings ranging from A (outstanding) to E (substandard). In addition, the spatial distribution of these letter ratings for each component and the total score was provided.

Liu found that among the 65 largest SMSAs, Portland, OR, ranked first in quality of life and Jersey City, NJ, last (Table 13). Among the 83 medium-size SMSAs, Eugene, OR, was rated number one and Mobile, AL, last. Among the 95 smaller SMSAs, LaCrosse, WI, placed first, and Laredo, TX, last. Geographically, the majority of large SMSAs with outstanding quality of life were found in the western region of the U.S. The substandard metropolitan areas were concentrated in the South and Northeast (Figure 14A). For medium-sized cities, the highest rated places were concentrated in the Midwest: Michigan, Wisconsin, and Indiana. The lowest rated medium-size metropolitan areas were again spatially concentrated in the South (Figure 14B). Finally, the highest quality of life in smaller SMSAs was found in upper Great Plains states, and in western New England. Substandard quality of life in smaller metropolitan regions was also concentrated in the South, East Texas and the Northeast (Figure 14C).

One of the strengths of Liu's study, in addition to its methodological rigor, was his use of an environmental indicator. To date, most quality-of-life studies have omitted an environmental component because of the paucity of data. Liu was able, through a survey of municipal and governmental officials and analysis of non-census data, to compile fairly detailed information on environmental quality within and among the 243 places. Liu evaluated environmental advantages and problems, including environmental amenities such as parks and recreation, as well as pollution indicators (air, noise, water, and solid waste). For this reason, Liu's analysis is one of the premier works in the field and one which has been often cited, used, and replicated.

The majority of these replications address specific issues, such as the quality of life in small-size cities, variations in the social quality of life among medium metropolitan areas or the role of quality of life in migration (Liu 1977, 1978; Hsieh and Liu 1983). In addition, each component of quality of life (e.g. environmental quality, political quality, social quality) has been re-analyzed for each group of city sizes.

One study in particular focussed on environmental quality in large metropolitan regions (Liu 1982). Although the data were 12 years old, Liu used a different methodology in constructing environmental quality indices and then compared these to his earlier work (Liu 1975a). Environmental quality was again measured using environmental amenity indicators (recreation) as well as pollution indicators (air, visual, noise, water, and solid waste). Liu found that the environmental quality of life was highest in the Pacific region, while many SMSAs located in the East were ranked near or at the bottom. Sacramento, California, had the best environmental quality according to Liu, while Chicago had the worst. It should be cautioned that this analysis used environmental quality data that are 12 years old; if one were to re-do the analysis using 1980 or later data, the rankings would change dramatically. For example 5 of Liu's top-10 metro regions (San Bernardino-Riverside-Ontario, Anaheim-Santa Ana-Garden Grove, Los Angeles-Long Beach, San Diego, and Phoenix) have severe air pollution problems; in 1982 they were ranked among the 10 worst places in the country for poor air quality (Council on Environmental Quality 1982).

Places Rated Almanac

In 1981, Boyer and Savageau published their *Places Rated Almanac;* it immediately became a best seller. The *Almanac* combined statistical information, place profiles, and anecdotal information on 277 metropolitan areas in the U.S. Nine subindices were created using 53 variables. These subindices included measures of climate, housing, health care and environment, crime, transportation, education, recreation, arts, and economics. A composite score on each indicator was computed for each of the 277 metropolitan regions which were then ranked from "best" to "worst." To provide an overall score, the ranking for each indicator for each metro region was summed. The lower the overall sum, the better the quality of life in that city.

The top-rated metro area in the country was Atlanta followed by Washington, DC, Greensboro-Winston-Salem-High Point, NC, Pittsburgh, PA, and Seattle-Everett, WA. In contrast, the lowest-rated metropolitan regions in the country were Lawrence-Haverhill, MA, (number 277) followed by Fitchburg-Leominster, MA, Pine Bluff, AR, Lowell, MA, and Panama City, FL (Table 14). The "best" places to live were concentrated in the Mid-South region in Tennessee, North Carolina, and Georgia (Figure 15A), a sharp contrast to Liu's study where no southern city reached the top ten. The "worst" places to live, according to *Places Rated Almanac,* were concentrated in the industrialized cities of New England and the agricultural cities of California's Central Valley. Again, this is in sharp contrast to the findings by Liu.

Because of the correlation between city size and quality of life and the preponderance of large cities with lower ratings, Boyer and Savageau provided a brief synopsis of quality of life in different size metro regions (Table 15). The top-rated small metro area (population under 250,000) was Asheville, NC, followed by Wheeling, WV,

TABLE 14   "BEST" AND "WORST" METRO AREAS, 1981

Top Twenty-five		Bottom Twenty-five	
1	Atlanta, GA	277	Lawrence-Haverhill, MA
2	Washington DC	276	Fitchburg-Leominster, MA
3	Greensboro-Winston-	275	Pine Bluff, AR
	Salem-High Point, NC	274	Lowell-Nashua, NH
4	Pittsburgh, PA	273	Panama City, FL
5	Seattle-Everett, WA	272	Fresno, CA
6	Philadelphia, PA	271	Bristol, CT
7	Syracuse, NY	270	Greeley, CO
8	Portland, OR	269	Lawton, OK
9	Raleigh-Durham, NC	268	Texarkana, TX
10	Dallas-Ft. Worth, TX	267	Vineland-Millville-
11	Knoxville, TN		Bridgeton, NJ
12	Nashville-Davidson, TN	266	Waterbury, CT
13	Anaheim-Santa-Ana-	265	Rockford, IL
	Garden Grove, CA	264	Macon, GA
14	Cleveland, OH	263	Sherman-Denison, TX
	San Francisco-Oakland, CA	262	Decatur, IL
16	Denver-Boulder, CO	261	Brockton, MA
17	Cincinnati, OH	260	New Britain, CT
18	Boston, MA		Stockton, CA
19	Louisville, KY	258	Modesto, CA
20	Miami, FL	257	Meriden, CT
21	Chicago, IL	256	Lafayette, LA
22	San Diego, CA	255	Ft. Smith, AR
23	Minneapolis-St. Paul, MN	254	Killeen-Temple, TX
24	St. Louis, MO	253	Paterson-Clifton-
25	Utica-Rome, NY		Passaic, NJ

Data from Boyer and Savageau 1981.

and Stamford, CT. In medium-size metro regions (population 250,000-1 million) the best place to live was Greensboro-Winston-Salem-High Point, NC, followed by Syracuse, NY, and Raleigh-Durham, NC. Finally, in large metro areas (population over 1 million) the highest-ranked city was Atlanta followed by Washington, DC, and Pittsburgh, PA.

Although the *Places Rated Almanac* received widespread media attention, it was not viewed as a critical success by the research community. Unlike Liu's work which was replicated, there have been relatively few studies which critique or build upon the work by Boyer and Savageau. One such study was presented at the annual meetings of the Association of American Geographers in 1984. Drawing on the data provided by Boyer and Savageau, Pierce (1984) recalculated the overall scores. This recalculation changed the rankings, leading Pierce to assert that there was a disparity between the perceived image of a city and its ranking in the *Places Rated Almanac*.

In order to test this assertion, Pierce asked a sample population of 1,100 New York state residents to rank the nine criteria used by Boyer and Savageau according to their importance in determining where to live. The highest rated criteria included economics, climate and crime, while the lowest were recreation (#7), transportation (#8) and the arts (#9). These rankings were then used to weight the nine aspects of urban places.

TABLE 15   "BEST" RANKINGS BY METRO SIZE, 1981

Rank	Small[a]	Medium[b]	Large[c]
1	Asheville, NC	Greensboro-Winston-Salem-High Point, NC	Atlanta, GA
2	Wheeling, WV	Syracuse, NY	Washington DC
3	Stamford, CT	Raleigh-Durham, NC	Pittsburgh, PA
4	La Crosse, WI	Knoxville, TN	Seattle-Everett, WA
5	Green Bay, WI	Nashville-Davidson, TN	Philadelphia, PA
6	Lincoln, NE	Louisville, KY	Portland, OR
7	Roanoke, VA	Utica-Rome, NY	Dallas-Ft. Worth, TX
8	Amarillo, TX	Honolulu, HI	Anaheim-Santa Ana-Garden Grove, CA
9	Sioux Falls, SD	Richmond, VA	Cleveland, OH
10	Galveston-Texas City, TX	Tucson, AZ	San Francisco-Oakland, CA

[a]Population under 250,000
[b]Population 250,000-1 million
[c]Population over 1 million
Data from Boyer and Savageau 1981.

For example, the rankings on the economic measure were multiplied by a factor of nine, while the rankings on the arts measure were multiplied by a factor of one. The products were then summed for each urban place, and the places were subsequently re-ranked.

Pierce found a positive association between his weighted ranking scheme and the unweighted one employed by Boyer and Savageau. He also found a concentration of top-rated cities in the South. In Pierce's ranking scheme, Greensboro, NC, ranked first followed by Knoxville, TN, Asheville, NC, Nashville, TN, and Raleigh, NC (Table 16). Only three cities, Atlanta, Greensboro, and Raleigh, appear on both Pierce's and Boyer and Savageau's top ten list. The "worst" cities, according to Pierce, were Fresno, CA, Lawrence-Fitchburg, MA, Lawton, OK, and Stockton, CA. Six of these also appear on Boyer and Savageau's bottom ten list. Spatially, the "best" places to live are concentrated in the coastal valleys of the Pacific Coast states and in the interior valleys of the Southeast (Figure 15B). The "worst" places to live are concentrated in the mill towns of New England, the agricultural communities in California's Central Valley, and the industrial cities near Chicago.

Pierce's attempt to incorporate a perceptual component in the weighting of quality of life components was a good one. His study showed, at least initially, the role and value of perceptual measures to weight objective social and environmental indicators in determining the overall quality of life of metropolitan areas.

In early 1985, Boyer and Savageau published a revised and updated *Places Rated Almanac*. The revision expanded the number of metropolitan regions to 329, altered and updated some variables, and changed the calculation of at least one subindex. Aside from these, the methodology including the 9 subindices is essentially unchanged from the earlier volume.

The top-three places to live in 1985 were Pittsburgh, PA, Boston, MA, and Raleigh-Durham, NC (Table 17). Atlanta, tops in 1981, dropped to 11th place in 1985, while Washington, DC, dropped from second to 15th place and Greensboro-Winston-Salem-High Point, NC, from third to 41st. The "worst" places to life included Yuba City,

CA, Pine Bluff, AR, and Modesto, CA (Table 17). The "worst" place to live in 1981, Lawrence-Haverhill, MA, improved its standing in four years and now rates a respectable 154. Spatially, the "best" places to live were concentrated in the Northeast urban corridor while the "worst" places were located in the South extending to Texas, the industrialized Midwest, and the northern Central Valley of California (Figure 15C). This contrasted quite sharply with the spatial distribution of "worst" places in the 1981 volume (Figure 15A). The rankings according to metropolitan size are also quite different (Table 18).

Following the publication of the newly revised *Places Rates Almanac,* Pierce (1985) also updated his earlier rankings. Using the same system of weights derived from his earlier study (1984), Pierce found Nassau-Suffolk, NY the most livable metropolitan area followed by Raleigh-Durham, NC and Norwalk, CT. Pittsburgh, the top-rated place according to Boyer and Savageau (1985) dropped to sixth place in Pierce's study. The lowest-rated metropolitan area, according to Pierce, was Pine Bluff, AR followed by Flint, MI and Rockford, IL. Yuba City, the "worst" metro area in *Places Rated Almanac* improved its position in Pierce's study to 326.

Comparing the findings of the two *Places Rated Almanac* studies is difficult because of a number of changes in the revised edition. The first involves changes in the geographic boundaries of metropolitan areas. In response to a governmental review of

TABLE 16 "BEST" AND "WORST" METRO AREAS, 1984

Top Twenty-five		Bottom Twenty-five	
1	Greensboro, NC	277	Fresno, CA
2	Knoxville, TN	276	Lawrence, MA
3	Asheville, NC	275	Fitchburg, MA
4	Nashville, TN	274	Lawton, OK
5	Raleigh, NC	273	Stockton, CA
6	Charleston, WV	272	Pine Bluff, AR
7	Wheeling, WV	271	Texarkana, TX
8	Evansville, IN	270	Rockford, IL
9	Anaheim, CA	269	Lowell, MA
10	Atlanta, GA	268	Paterson, NJ
11	San Jose, CA	267	Great Falls, MT
12	Galveston, TX	266	Bakersfield, CA
13	Portland, OR	265	Jackson, MI
14	Louisville, KY	264	Macon, GA
15	Seattle, WA	263	St. Joseph, MO
16	Lexington, KY	262	Lewiston, ME
17	Eugene, OR	261	Bristol, CT
18	Fort Wayne, IN	260	New Britain, CT
19	Washington DC	259	Gary, IN
20	Huntington, WV	258	Meriden, CT
21	New Brunswick, NJ	257	Waterbury, CT
22	Pittsburgh, PA	256	Peoria, IL
23	Salem, OR	255	Kankakee, IL
24	Utica, NY	254	Corpus Christi, TX
25	Johnstown, PA	253	Pittsfield, MA

Data from Pierce 1984.

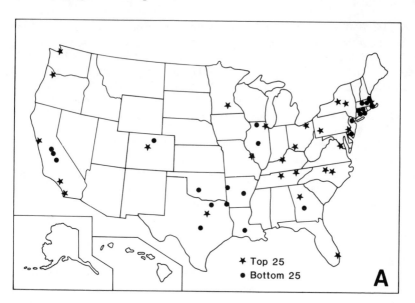

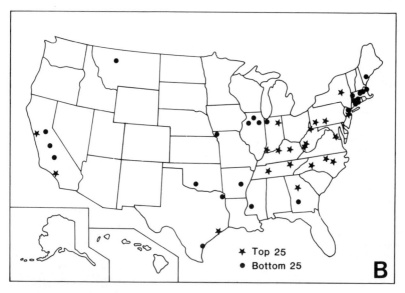

FIGURE 15   RATING PLACES, 1981 AND 1985. Metropolitan rankings for **A,** 1981 (based on Boyer and Savageau, 1981); **B,** 1981 re-ranked by Pierce (1984); **C,** 1985 (Boyer and Savageau, 1985).

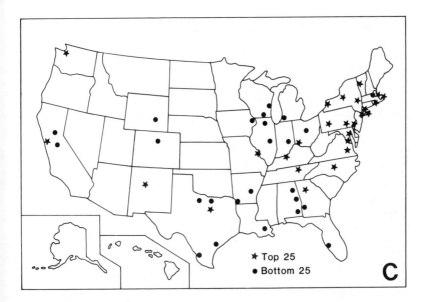

Top 25 ★
Bottom 25 ●

metropolitan definitions in 1983, a new system of metropolitan designation was implemented. The former designation of SMSA is now defined as either a Metropolitan Statistical Area (MSA) or a Primary Metropolitan Statistical Area (PMSA). Only 166 of the 329 areas studied in 1985 have the same geographic boundaries as in 1981. The remainder had either added or deleted counties, merged with other metropolitan areas, or were entirely new areas as a result of increased urban growth. In addition, where there were two or more urban areas adjacent to one another, a metropolitan complex is designated and termed Consolidated Metropolitan Statistical Area (CMSA). An example of a CMSA is the New York-Northern New Jersey-Long Island CMSA which includes 12 PMSAs. While the study focussed on MSAs and PMSAs, the contribution of CMSAs to the overall ranking is significant.

Another important factor influencing lack of comparability was a change or modification of criteria used to compile each subindex. If the same measures had been used in both studies, more recent data might influence the ultimate ratings, but would not reduce the comparability among the two studies. There were, however, a number of changes in the weighting or scoring of factors which influence the subindex. Changes in the variables used to construct the health care and recreation subindices prohibits comparisons from 1981 to 1985. One of the most significant changes — which not only influences comparability, but also the absolute ranking of places — is the question of urban access. Certain attributes of urban areas such as a professional sports team, international airport, or symphony may be located in one metropolitan area, yet are enjoyed by people living in another. In other words, Nassau-Suffolk, NY, is awarded bonus points, thus improving its ratings because of its proximity to New York City which has some of these amenities. This promixity factor is labeled CMSA Access and strongly influences the rankings of metro areas particularly in the health care, transportation, cultural, and recreational subindices.

TABLE 17   "BEST" AND "WORST" RATED PLACES, 1985

Top Twenty-five		Bottom Twenty-five	
1	Pittsburgh, PA	329	Yuba City, CA
2	Boston, MA	328	Pine Bluff, AR
3	Raleigh-Durham, NC	327	Modesto, CA
4	San Francisco, CA	326	Dothan, AL
5	Philadelphia, PA-NJ	325	Albany, GA
6	Nassau-Suffolk, NY	324	Benton Harbor, MI
7	St. Louis, MO-IL	323	Gadsden, AL
8	Louisville, KY-IN	322	Casper, WY
9	Norwalk, CT	321	Rockford, IL
10	Seattle, WA	320	Anderson, IN
11	Atlanta, GA	319	Victoria, TX
12	Dallas, TX	317	Laredo, TX
13	Buffalo, NY		Texarkana, TX-AR
14	Knoxville, TN	316	Fitchburg-Leominster, MA
15	Baltimore, MD	315	Anniston, AL
	Washington, DC-MD-VA	314	Sheboygan, WI
17	Cincinnati, OH-KY-IN	313	Mansfield, OH
18	Burlington, VT	312	Greeley, CO
19	Albany-Schenectady-Troy, NY	311	Baton Rouge, LA
20	Syracuse, NY	310	Bradenton, FL
21	Albuquerque, NM	309	Decatur, IL
22	Harrisburg-Lebanon-Carlisle, PA	308	Sherman-Denison, TX
23	Richmond-Petersburg, VA	307	Janesville-Beloit, WI
24	Providence, RI	306	Wichita Falls, TX
25	New York, NY	305	Dubuque, IA

Data from Boyer and Savageau 1985.

As a result of these methodological and data changes, the revised edition of *Places Rated Almanac* is not a longitudinal study of "best" and "worst" places that can be compared to the previous work, but an entirely new study. Although the temptation may be great, drawing comparisons between 1981 and 1985 rankings is inappropriate.

Other Inter-Urban Studies: "The Worst American City"

Although we have discussed the major inter-urban quality of life studies, there are three minor studies that are worth noting. While these are not as significant as those by Liu, Thorndike, or Boyer and Savageau, they are nevertheless briefly described here because either they appeared in popular magazines or were written by geographers.

In January 1975 Arthur M. Louis published an article in *Harper's Magazine* titled "The Worst American City." This article used Angoff and Mencken's 1931 study of American states as its model. Louis' primary purpose was to provide a "scientific study to confirm or deny your prejudices." As Louis later stated in his reasoning for the title of the article (1975:67):

> I'd like to believe that it doesn't require any perversity of character to go looking for the worst city. There are no good cities in America today — only bad and less bad.

TABLE 18   "BEST" RANKINGS BY METRO SIZE, 1985

Rank	Small[a]	Medium[b]	Large[c]
1	Norwalk, CT	Raleigh-Durham, NC	Pittsburgh, PA
2	Burlington, VT	Louisville, KY	Boston, MA
3	Charlottesville, VA	Knoxville, TN	San Francisco, CA
4	Asheville, NC	Albany-Schenectady-Troy, NY	Philadelphia, PA-NJ
5	Stamford, CT	Syracuse, NY	Nassau-Suffolk, NY
6	Portland, ME	Albuquerque, NM	St. Louis, MO-IL
7	Danbury, CT	Harrisburg-Lebanon-Carlisle, PA	Seattle, WA
8	Galveston,-Texas City, TX	Richmond-Petersburg, VA	Atlanta, GA
9	South Bend-Mishawaka, IN	Providence, RI	Dallas, TX
10	Middletown, CT	Middlesex-Somerset-Hunterdon, NJ	Buffalo, NY

[a]Population under 250,000
[b]Population 250,000 and 1 million
[c]Population over 1 million
Data from Boyer and Savageau 1985.

Louis studied the nation's fifty largest cities and included 24 variables in his analysis. These variables were used to construct crime, health, affluence, housing, education, atmosphere, and amenities subindices. The simple linear additive methodology followed that of Mencken — summing of the rankings for each subindex for each city, then re-ranking for the overall score.

Seattle, Tulsa, San Diego, San Jose, and Honolulu were the top-rated cities, while Baltimore, Detroit, Chicago, St. Louis, and Newark, NJ, round out the bottom (Table 19). In his conclusions, Louis mentioned that these rankings do not include any weighting of the factors, nor do they measure the vitality or excitement of cities which might influence people's image of them. Louis's study, like that of Angoff and Mencken, confirmed what most people think. His article did reach a wide audience and added information that either confirmed people's images of cities or refuted them. In remarking about Newark's low finish, Louis wrote (1975:71):

> The city of Newark stands without serious challenge as the worst of all. It
> ranked among the worst five cities in no fewer than nineteen of the
> twenty-four categories, and it was dead last in nine of them. . . . Newark
> is a city that desperately needs help.

Most of the inter-urban quality of life studies to date were conducted by non-geographers, although within the last few years, urban geographers have shown an increasing interest in the field. In 1981, for example, there was an urban geography session at the annual meetings of the Association of American Geographers which featured two papers on quality of life. Dakan (1981) examined medium-sized SMSAs and classified them into overall quality of life groupings. He found no significant difference in ratings between northern and southern cities once the racial composition of the city was held constant. Miller's paper (1981) derived a simplified index of quality

TABLE 19   LARGE CITY RATINGS BY LOUIS (1975)

Top Ten		Bottom Ten	
1	Seattle	50	Newark
2	Tulsa	49	St. Louis
3	San Diego	48	Chicago
4	San Jose	47	Detroit
5	Honolulu	46	Baltimore
6	Portland	45	Birmingham
7	Denver	44	Jacksonville
8	Minneapolis	43	Cleveland
9	Oklahoma City	42	Norfolk
10	Omaha	41	San Antonio

Data from Louis 1975.

of life using seven sub-categories including housing, social disorganization, education, economic status, amenities, health and crime. He used the fifty largest cities in the U.S. in order to compare his results to those of Louis (1975) and Liu (1975a). Miller found a high degree of correspondence between his ratings and those of earlier studies. In the Miller study, Minneapolis-St. Paul was ranked first, while Birmingham, AL, was last.

Quality of Life Within Cities: The Intra-Urban Scale

There are very few studies of quality of life at the intra-urban scale. Those that have been conducted use only social indicators and methodologically follow the social-area analysis and territorial-indicator models described in Chapter 2. Curiously, geographical studies of quality of life are noticeably absent at other scales, whereas they dominate at the intra-urban level. Three studies are worth noting, all of which focussed on southern cities, Gainesville, FL (Dickinson et al. 1982), Atlanta, GA (Bederman 1974), and Tampa, FL (Smith 1973). All three represent case studies aimed at developing a set of intra-city social indicators. These could then be used to identify areas where assistance is needed in order to improve the quality of life of the area's residents. In this respect, these studies represent updated versions of the earlier factorial ecologies and social-area analyses conducted by both geographers and sociologists. The findings are in fact, quite similar. The results of the factorial ecologies conclude that urban residential areas are differentiated on the basis of three factors: socio-economic status, life cycle, and ethnic composition or racial segregation. This was also true to some extent for Gainesville and Tampa, although social problems and social deprivation factors were more important in distinguishing areas. This finding is a function of the use different variables, many of which are available at the local level only.

The results of these intra-urban studies suggest that their primary use is not so much in determining and comparing places to live, but rather in describing places (neighborhoods) where wide disparities exist in social conditions and social well-being. In all three studies, the authors argued for the merits of this type of analysis in planning decisions and allocation of urban services. However, few, if any, intra-urban quality of life studies in the U.S. have been done since. In Britain, territorial social indicators continues as a viable research area in human geography and urban planning (Pacione 1982). Intra-urban analyses of quality of life are the least successful in differentiating places, as the findings are not significantly different from research conducted twenty-five years ago. While perceptual indicators of quality of life could be employed, none have. In addition, measures of environmental quality would be difficult to obtain at such a small scale without actual monitoring or collection of raw data. This is not to suggest that community and neighborhood analyses are not useful; they are. Rather, it suggests that quality of life is a broader concept and has its most appropriate application at the inter-city level, or larger scale.

Summary

The studies reviewed here employ a wide range of methodologies to derive rankings ranging from simple linear addition techniques to more sophisticated multivariate analyses. All, however, generate headlines regardless of how good or bad they are from a statistical or methodological perspective. Ranking studies have become so prevalent and subject to such controversy, that we might even consider rating places a national pastime. As we shall see in the next chapter, many problems arise when these ratings of places are viewed as fact and used indiscriminately by scholars, the media, or the public to rank the "best" and "worst" places to live, work, or play. While Thorndike cautioned people "not to take anyone's opinion about your city, get the facts" (1939:150), Angoff and Mencken were perhaps more prophetic (1931:2):

Statistics, to be sure, are not always reliable, but we have nothing better, and we must make as much of them as we can.

4

Why Do People Care About Quality of Life?

Whether the subject is the beefiest burger or the biggest corporation,
Americans have a penchant for making lists of the best and the worst,
then arguing about the results. . . . No rankings have inspired more
disagreement than those about home sweet home.
— *Time Magazine,* March 11, 1985

Rating places is big business. Not only do national magazines such as *MS* and *Money* evaluate and rank places for their specialized audiences, but industry, real estate, and marketing firms conduct internal evaluations of places to aid in locational or marketing decisions. The commercial success of *Places Rated Almanac* prompted a specialized guide on retirement living which ranks and compares 107 counties in the U.S. (Boyer and Savageau 1983). The "best" retirement county? Brevard, North Carolina; the "worst?" Lanconia-Lake Winnipesaukee, New Hampshire. This popular interest in quality of life reflects an eagerness on the part of the public for practical and reliable information about places. While governmental data sources are available to the general public, rarely do people have the time, inclination, or skills to gather and analyze the information. It is much easier to run to a local bookstore and purchase place-rating guides.

Aside from popular appeal, there is also another level of interest in quality of life by the academic and public policy communities. This involves the use of quality of life indicators to measure social improvement and explain changes in society. This chapter examines the sometimes conflicting notions of place and the quality of life found there. We will look at how quality of life studies are used, what can be learned from them, how to improve them, and public reactions to rating places.

Quality of Life and Public Policy

Quality of life research is used in two different public policy areas: monitoring public policy and locational decisions. This public policy research involving quality of life rarely receives media attention. Its objective is to evaluate and compare attributes among places, not to rank them from "best" to "worst."

Public Policy Monitoring

Early quality of life studies monitored social and environmental progress and changes over time. These studies were conducted almost exclusively by sociologists who only examined the social environment and focussed on the individual and how the individual fared in society at large. Only later did geographers incorporate the notion of space into these monitoring programs, introducing the concept of "territorial social indicators" which documented spatial variations in social condition.

The vast majority of current quality of life research simply continues to monitor objective conditions (social, environmental) of areas (neighborhoods, cities, states, countries) and the spatial distribution of these objective conditions. These spatial comparisons are used to assess the success or failure of government programs, particularly in relation to the resources allocated to them. Studies are designed to highlight inequities in the distribution of resources by examining spatial variations in individual quality of life. Such an approach permits decision makers to monitor the success of their programs and to reallocate resources to needed areas to achieve desired results. Again, the vast majority of this work examines urban and intraurban spatial variations in social conditions. In this respect, quality of life research provides data for public policy decision making.

On another level, quality of life analyses are used to suggest a national agenda for improving quality of life, including progress toward quantitative and qualitative policy goals such as decreasing poverty and improving public education (President's Commission 1980). Similar quality of life studies also provide the basis for comparative assessments of well-being at the international level (Szalai and Andrews 1981). Public policy uses of quality of life research range from simple data collection to inputs into policy decisions, the latter ranging from resource allocations and new program development to regulatory compliance.

Locational Decision Making

A second major use of quality of life research is to differentiate and evaluate places in order to explain social and demographic changes at both the regional and national levels. Economists use this form of quality of life research to explain economic and locational behavior of individuals and firms, especially the profit-maximizing firms (Wingo and Evans 1977; Katzner 1979; Power 1980). These studies generally address labor markets and wage differentials. For example, an industry located in a degraded area (implying a lower quality of life) must pay higher wages to entice employees compared to firms in higher quality of life regions with lower wages but more amenities. If true, such correlations between quality of life and rising labor costs results in drastic spatial changes in regional economies prompting mass migration of both industry and people.

Migration decisions are nominally based on perceived and real employment opportunities. There is increasing empirical evidence, however, that quality of life is also important, especially when two places are equally attractive from an economic standpoint (Graves 1973, 1980; Liu 1975b; Porell 1982; Hsieh and Liu 1983). For example, in examining intermetropolitan migration rates from 1965-70, Porell (1982) found that differences in quality of life (measured by Liu's [1975a] variables) not only existed, but apparently influenced the volume of migration into that area. Those metropolitan areas with a good quality of life had a significant competitive advantage over others in attracting new residents despite small differentials in economic opportunity. Such researcn explains the behavior of prospective employers who use quality of life to induce new employees while paying lower wages. The economic choice is clear: a better quality of life and lower wages, or a low quality of life and higher wages. Hsieh and Liu (1983) found a similar result. Their examination of inter-regional migration and regional differences in quality of life found that the short-run pursuit of better environmental quality was the most important factor in migration. In the long run, the pursuit of a better social quality of life influenced the migration decision. While the role of quality of life in influencing inter-regional migrations is documented, intra-urban migrations (suburbs to cities) are poorly explained by the concept (Roseman 1976).

Improving Quality of Life Studies

As mentioned in Chapter 2, there are many criticisms of quality of life studies ranging from poor, inadequate, and inappropriate data to problems with measurement techniques and statistical analyses. All too often, conclusions are made which are not necessarily drawn from the data. Ecological inference is the most common methodological problem. The ecological fallacy occurs when an average characteristic of an area is ascribed to all the inhabitants in that area. For example, if the mean income of a county is $11,000, then everyone in that county is assumed to earn that amount for statistical purposes. As most of the social indicators used in quality of life studies are from the U.S. Bureau of the Census, which uses the census tract as the primary unit of measurement, we may unknowingly be committing this mistake from the start.

The ranking techniques used in many quality of life studies can be statistically manipulated to produce different results. For example, Smith (1973) ranked U.S. states using two different statistical techniques (additive and principal component models). In the additive model, the top ranked state was Connecticut, which placed third using the other technique (Table 10, Chapter 3). Mississippi, on the other hand, ranked 48th using both models. Using Angoff and Mencken's ranking technique applied to 1980 data, we find that Connecticut remains in the first position while Louisiana ranks last (Table 9, Chapter 3). If we were to average the rankings for each subindex (to account for missing data), total each subindex, compute the average, and then rank these averages, a different hierarchy of states emerges. Using this very simple technique, Wyoming tops the list while Connecticut drops to sixth place. Louisiana, however, still remains at the bottom. In short, it is very easy to massage numbers to obtain the results the researcher seeks.

A satisfactory quality of life study is difficult to conduct because of the type of data that are necessary and our limited understanding of the relationships between the various components. There are, however, a number of guidelines, or points to ponder, that may help in evaluating quality of life research.

(1)   Quality of life studies should incorporate measures of all three components (social, environmental, perceptual) not just one (see Figure 1 in Chapter 1).

(2)   Analytical methods must include measurements of both objective data on social and environmental conditions and subjective data on individual appraisals of these conditions. The method should also incorporate subjective assessments of the relative importance of variables to individuals, as well as individual images of, and attachments to, place. Objective data are readily available from published sources, while subjective data are not. Thus subjective measures are absent in most quality of life research.

(3)   The indicators chosen must actually measure the desired concept or attribute. When looking for an indicator of environmental quality, for example, one could choose the average number of days over 80 degrees Fahrenheit or the number of golf courses. One is a better measure of physical comfort, while the other is a better indicator of recreational amenities. Which one is better? It depends on the intent of researcher, and the questions posed.

(4)   The scale of data must be specified and different scales should not be mixed (census tract with block data or incorporated city *vs.* metropolitan area). The data for an entire metropolitan area, for example, might reflect the pattern of very wealthy suburbs with an old, economically depressed central city. Statistics collected for entire metropolitan regions and then attributed to a particular city give an inaccurate picture of conditions. This problem is particularly acute because of the nature of available statistics. Most people recognize Chicago, for example, as a single unit, the incorporated city. Available statistics, particularly those compiled by governmental sources, variously refer to Chicago (the incorporated city), Chicago (the metropolitan area including Cook, Du Page and McHenry counties), or Chicago (the megalopolis stretching north to Kenosha, WI; east to Gary-Hammond, IN; and south and west to Lake County, Joliet, and Aurora-Elgin, IL).

(5)   The objective of the research should clearly state whether the focus is on individual quality of life within an area or spatial variations in quality of life between areas. What an individual regards as a good quality of life may have nothing to do with a particular place. The quality of life of a place, however, is intrinsically linked not only to a specific locale but to the people who live and work there.

(6)   Statistical overkill does not always produce the best results, nor does it imply the best study. While sophisticated statistical techniques are useful in reducing large data bases, sometimes a more simplified approach is warranted and produces similar results.

(7)   No matter how well done, someone will not like nor appreciate the rank attributed to their hometown, state, or country. By definition, place rankings spring from subjective preferences, for which everyone has a different set. The most satisfactory type of quality of life study is one where data (mostly derived from published sources) are presented covering a variety of attributes of places and ranked from highest to lowest or "best" to "worst." Individuals, using these objective indicators, then incorporate their own set of preferences on the relative importance of each to their overall quality of life. People's true preferences would come to life (economic conditions versus environmental amenities) and this system would allow sensible tradeoffs between them. For example, Sally likes the climate and cultural amenities of Los Angeles, Steve likes the environmental amenities of Seattle; they compromise and move to San Francisco. Avoid rankings as much as possible unless you are very explicit on how these were derived, and be prepared for public scrutiny.

FIGURE 16   *PLACES RATED ALMANAC* AND MEDIA ATTENTION.
Controversy abounds in ranking places from "best" to
"worst."

Place Pride, Boosterism, and Chauvinism

The fascination with lists and rankings, coupled with the potential role of quality of
life in locational decisions of both individual and industry, partially explains the public
interest in rating places. Place boosterism and civic pride can improve a city's, town's,
state's, or country's image, but a good ranking brings national exposure and free
publicity (Figure 16). Conversely, a poor showing agitates civic leaders to such an

TABLE 20   REACTIONS TO 1985 RANKINGS IN THE *PLACES RATED ALMANAC*

"It's not that we don't deserve to be No. 1. It's just that we're simply not used to being on top in anything that doesn't involve football. Now we have every reason to fear a Yuppie invasion. As you know, Yuppies take lists and ratings very seriously." — Peter Leo, columnist, *The Pittsburgh Post-Gazette*.

"Yes, it is a good place if you are earning more than $25,000 a year, love winter and adore potholes." — Letter to *Time Magazine* from a Pittsburgh resident.

"We're the best kept secret in the world and now the secret is out." — Pittsburgh Mayor Richard Caliguiri.

"Pittsburgh is kind of like Newark without the cultural advantages." — Johnny Carson.

"We're planning a little get-together on the 10th Street bridge to burn Rand McNally maps." — Ron Haedicke, county fair manager, Yuba City, CA.

"Yuba City isn't evil. It isn't bad. It's just not very much." — Richard Boyer, co-author of *Places Rated Almanac*.

"We capitalize on it. We mentioned it in ads placed in different magazines and periodicals around the country." — President of the Greensboro, NC Chamber of Commerce.

"These unsolicited surveys that are done by somebody because they want to do a survey and release it publicly can be absolutely devastating to a city's economy." — Tulsa, OK Mayor, Terry Young.

"I've lost about 2 weeks of my life returning telephone calls to irate citizens. I think it should be illegal for this type of analysis to be completed." — Fayetteville, NC Mayor, Bill Hurley.

"It's rather astonishing the degree of interest this type of research generates." — Robert Pierce, geographer at SUNY Cortland.

"We get a lot of mail about our book. Some of it ticking." — Richard Boyer and David Savageau, authors. *Places Rated Almanac*.

Compiled from *USA Today* February 28, 1985; *New York Times* March 31, 1985; and *Time Magazine* March 11 and April 8, 1985.

extent that they may seek legal restitution. The city of Tulsa filed a $26 million lawsuit against a political science professor at Cleveland State University for defaming the city's reputation (*USA Today* 1985). The study placed Tulsa last in two of 10 categories. To add insult to injury, the professor acknowledged that some of his data were wrong, apologized, and re-ranked the city, but the city officials were not satisfied. They estimated they will spend more than $2 million in publicity (up from $700,000 the previous year) as a result of the damaging effects of the study. The suit was eventually dropped (*Chronicle of Higher Education* 1985). Fortunately, not everyone has such an extreme reaction to place rankings (Table 20).

Governmental and quasi-governmental groups such as chambers of commerce, civic associations, local business groups, and tourist boards recognize not only the importance of a city's or state's image to tourists but also to new residents and industry. When a city, such as Yuba City, CA, is rated as "worst" in the country, it not only wounds civic pride, but disenchants potential new residents and industries.

Considerable amounts of money have been spent over the decades to upgrade images of place and to attract new residents and, more importantly, new industry. All of this illustrates the lengths to which tourism boards, local civic organizations and chambers of commerce are willing to go in projecting a positive image of place. We see this place boosterism for tourist dollars constantly in ad campaigns such as New York's "I Love NY," and Virginia's "Virginia is for Lovers." Popular music often accompanies these ad campaigns on television and radio. Songs can either boast the merits of one place exclusively ("New York, New York," or "I Left My Heart in San Francisco") or highlight the image of one place at the expense of another. Randy Newman's popular "I Love LA," used as the promotional song by commercial sponsors of the 1984 Summer Olympic Games in Los Angeles, is one example:

Hate New York City
It's cold and it's damp
And all the people dress like monkeys
Let's leave Chicago to the Eskimos
That town's a little bit too rugged
For you and me, you bad girl

From the South Bay to the Valley
From the West Side to the East Side
Everybody's very happy
'Cause the sun is shining all the time
Looks like another perfect day

I Love L.A. (We love it)
I Love L.A. (We love it)

(Music and Words by Randy Newman, Copyright Six Pictures Music,
1983, used with permission)

No matter how empirically defined the quality of life of places becomes, there will always be a very personal element that will defy measurement. Just as beauty is in the eyes of the beholder, so may be a place's quality of life.

Bibliography

Abler, R.F., J.S. Adams, and P. Gould. 1971. *Spatial Organization: The Geographer's View of the World.* Englewood Cliffs, NJ: Prentice-Hall.

Abler, R.F. and J.S. Adams. 1976. *A Comparative Atlas of America's Great Cities: Twenty Metropolitan Regions.* Minneapolis: University of Minnesota Press.

Andrews, F.M. and S.B. Withey. 1976. *Social Indicators of Well-Being.* New York: Plenum.

Angoff, C. and H.L. Menken. 1931. "The Worst American State: Part I, II, III," *American Mercury* 24 (93-95):1-16, 175-188, 355-370.

Bauer, R. (editor). 1966. *Social Indicators.* Cambridge, MA: MIT Press.

Bayless, H. 1983. *The Best Towns in America.* Boston, MA: Houghton Mifflin.

Bederman, S.H. 1974. "The Stratification of Quality of Life in the Black Community of Atlanta, Georgia," *Southeastern Geographer* 14(1):26-37.

Bednarz, R.S. 1975. *The Effect of Air Pollution on Property Value in Chicago.* Chicago, IL: University of Chicago, Department of Geography, Research Paper 166.

Berry, B.J.L. and F.E. Horton. 1970. *Geographic Perspectives on Urban Systems.* Englewood Cliffs, NJ: Prentice-Hall.

Berry, B.J.L. and J.D. Kasarda. 1977. *Contemporary Human Ecology.* New York: Macmillan.

Berry, B.J.L. et al. 1977. *The Social Burdens of Environmental Pollution.* Cambridge, MA: Ballinger.

Bisselle, C.A. et al. 1972. *National Environmental Indices: Air Quality and Outdoor Recreation.* Washington, DC: MITRE Corp.

Boney, F.N. 1976. "The American South," *Journal of Popular Culture* 10:290-97.

Boulding, K. 1961. *The Image.* Ann Arbor, MI: University of Michigan Press.

Bowman, T.F. et al. 1981. *Finding Your Best Place to Live in America.* New York: Warner Books.

Boyer, R. and D. Savageau. 1981. *Places Rated Almanac.* Chicago, IL: Rand McNally.

Boyer, R. and D. Savageau. 1983. *Places Rated Retirement Guide.* Chicago, IL: Rand McNally.

Boyer, R. and D. Savageau. 1985. *Places Rated Almanac* (Revised Edition). Chicago, IL: Rand McNally.

Buttimer, A. and D. Seaman (editors). 1980. *The Human Experience of Space and Place.* New York: St. Martin's Press.

Campbell, A. and P.E. Converse. 1972. *Human Meaning of Social Change.* New York: Russell Sage Foundation and Basic Books.

Campbell, A. and W.L. Rodgers. 1976. *The Quality of American Life: Perceptions, Evaluations and Satisfactions.* New York: Russell Sage Foundation.

Canter, D. 1983. "The Purposive Evaluation of Places: A Facet Approach," *Environment and Behavior* 15:659-98.

Caris, S.L. 1978. *Community Attitudes Toward Pollution.* Chicago, IL: University of Chicago, Department of Geography, Research Paper 188.

Carley, M. 1981. *Social Measurement and Social Indicators.* Winchester, MA: Allen and Unwin.

Chronicle of Higher Education. 1985. July 10, 1985.

Conway, H. McK. and L.L. Liston. 1981. *The Good Life Index: How to Compare Quality of Life Throughout the U.S. and Around the World.* Atlanta, GA: Conway Publications.

Craik, K.H. and E.H. Zube (editors). 1976. *Perceiving Environmental Quality.* New York: Plenum.

Cutter, S.C. 1981. "Community Concern for Pollution: Social and Environmental Influences," *Environment and Behavior* 13 (1):105-24.

Cutter, S.L. et al. 1985. "Angoff and Mencken's 'Worst American State' Revisited: Interstate Rankings of Quality of Life," unpublished manuscript.

Dahmann, D.C. 1981. "Subjective Indicators of Neighborhood Quality," in D.F. Johnston (editor), *Measurement of Subjective Phenomena.* Washington DC: U.S. Department of Commerce, Bureau of the Census, Special Demographic Analyses (CDS-80-3).

Dakan, A.W. 1981. "Intra- and Inter-urban Variations in Quality of Life," Paper presented at the Annual Meetings, Association of American Geographers, Los Angeles, CA.

Dearden, P. 1980. "A Statistical Method for the Assessment of Visual Landscape Quality for Land-use Planning," *Journal of Environmental Management* 10:51-68.

Dickinson, J.C. III et al. 1972. "The 'Quality of Life' in Gainesville, Florida: An Application of Territorial Social Indicators," *Southeastern Geographer* 12(2):121-132.

Downs, R.M. and D. Stea. 1977. *Maps in Minds: Reflections on Cognitive Mapping.* New York: Harper and Row.

Duncan, O.D. 1969a. "Social Forecasting — The State of the Art," *Public Interest* 17:88-118.

Duncan, O.D. 1969b. *Towards Social Reporting: New Steps.* New York: Russell Sage Foundation.

Duncan, O.D. 1984. *Notes on Social Measurement, Historical and Critical.* Beverly Hills, CA: Sage Publications.

Elgin, D. et al. 1974. *City Size and the Quality of Life.* Washington, DC: National Science Foundation.

Estes, R.J. 1984. *The Social Progress of Nations.* New York: Praeger.

Garreau, J. 1981. *The Nine Nations of North America.* Boston, MA: Houghton-Mifflin.

Garwood, A.N. (editor). 1984. *199 American Cities Compared.* Burlington, VT: Information Publications.

Gould, P. and R. White. 1974. *Mental Maps.* Baltimore, MD: Pelican Books.

Graves, P.E. 1973. "A Reexamination of Migration, Economic Opportunity and Quality of Life," *Journal of Regional Science* 13:205-11.

Graves, P.E. 1980. "Migration and climate," *Journal of Regional Science* 20:227-238.

Gross, B.M. (editor). 1969. *Social Intelligence for America's Future: Explorations in Societal Problems.* Boston, MA: Allyn & Bacon.

Hartshorn, T. 1981. *Interpreting the City: An Urban Geography.* New York: Wiley.

Hawley, A.H. and O.D. Duncan. 1957. "Social Area Analysis: A Critical Appraisal," *Land Economics* 33:227-245.

Helburn, N. 1982. "Geography and the Quality of Life," *Annals, Association of American Geographers* 72:445-56.

Hill, A.D. et al. 1973. *The Quality of Life in America; Pollution, Poverty, Power and Fear.* New York: Holt, Rinehart and Winston.

Hsieh, C.T. and B.C. Liu. 1983. "Pursuance of Better Quality of Life: In the Long Run, Better Quality of Life is the Most Important Factor in Migration," *American Journal of Economics and Sociology* 42:431-40.

Hunter, A. 1974. *Symbolic Communities.* Chicago, IL: University of Chicago Press.

Inhaber, H. 1976. *Environmental Indices.* New York: Wiley.

Jackson, J.B. 1984. *Discovering the Vernacular Landscape.* New Haven, CT: Yale University Press.

Jacoby, L.R. 1972. *Perception of Air, Noise, and Water Pollution in Detroit.* Ann Arbor: University of Michigan Press.

James, P.E. 1972. *All Possible Worlds.* Indianapolis, IN: Odyssey Press.
Johnston, R.J. 1976. "Residential Area Characteristics: Research Methods for Identifying Urban Subareas — Social Area Analysis and Factorial Ecology," pp. 193-235 in D.T. Herbert and R.J. Johnston (editors), *Spatial Process and Form.* London: Wiley.
Katzner, D.W. 1979. *Choice and the Quality of Life.* Beverly Hills, CA: Sage Publications.
Kimball, T.L. 1972. "Why Environmental Quality Indices?" pp. 7-14 in Ott 1978.
King, L.J. 1969. *Statistical Analysis in Geography.* Englewood Cliffs, NJ: Prentice-Hall.
Knox, P.L. 1975. *Social Well-Being: A Spatial Perspective.* Oxford, UK: Clarendon Press.
Knox, P.L. 1978. "Territorial Social Indicators and Area Profiles," *Town Planning Review* 49: 75-83.
Kurian, G.T. 1979. *The Book of World Rankings.* New York: Facts on File.
Kurian, G.T. 1983. *The New American Gazetteer.* New York: Signet Books.
Kurian, G.T. 1984. *The New Book of World Rankings.* New York: Facts on File.
Land, K.C. 1983. "Social Indicators," *Annual Review of Sociology* 9:1-26.
Land, K.C. and S. Spilerman. 1975. *Social Indicator Models.* New York: Russell Sage Foundation.
Lave, L.B. and E.P. Seskin. 1977. *Air Pollution and Human Health.* Baltimore, MD: Johns Hopkins University Press for Resources for the Future.
Leopold, A. 1949. *Quantitative Comparisons of Some Aesthetic Factors among Rivers.* Washington, DC: U.S. Geological Survey Circular 620.
Ley, D. 1983. *A Social Geography of the City.* New York: Harper and Row.
Liu, B. 1975a. *Quality of Life Indicators in the United States Metropolitan Areas, 1970.* Washington, DC: Government Printing Office, U.S. Environmental Protection Agency.
Liu, B. 1975b. "Differential Net Migration Rates and the Quality of Life," *Review of Economics and Statistics* 57:329-337.
Liu, B. 1977. Economic and Non-economic Quality of Life: Empirical Indicators and Policy Implications for Large Standard Metropolitan Areas," *Journal of Economics and Sociology.* 36(3):225-40.
Liu, B. 1978. "Variations in Social Quality of Life Indicators in Medium Metropolitan Areas," *American Journal of Economics and Sociology* 37:241-60.
Liu, B. 1982. "Environmental Quality Indicators for Large Metropolitan Areas: A Factor Analysis," *Journal of Environmental Management* 14(2):127-38.
Louis, A. 1975. "The Worst American City: A Scientific Study to Confirm or Deny Your Prejudices," *Harper's Magazine* January:67-71.
Lowenthal, D. and M.J. Bowden (editors). 1976. *Geographies of the Mind.* New York: Oxford University Press.
Lukermann, F. 1964. "Geography as a Formal Intellectual Discipline and the Way in Which it Contributes to Human Knowledge," *Canadian Geographer* 8(4):167-172.
Lynch, K. 1960. *The Image of the City.* Cambridge, MA: MIT Press.
Lynch, K. 1972. *What Time is This Place?* Cambridge, MA: MIT Press.
McHarg, I. 1971. *Design with Nature.* Garden City, NY: Natural History Press for the American Museum of Natural History.
Marlin, J.T. and J. S. Avery. 1983. *The Book of American City Rankings.* New York: Facts on File.
Michener, J.A. 1970. *The Quality of Life.* Philadelphia, PA: Lippincott.
Miller, G.R. 1981. "The Quality of Life in America's Cities," Paper presented at the Annual Meetings of the Association of American Geographers, Los Angeles, CA.

Morris, M.D. 1979. *Measuring the Conditions of the World's Poor.* New York: Pergamon.

Murdie, R.A. 1969. *Factorial Ecology of Metropolitan Toronto, 1951-1961.* Chicago, IL: University of Chicago, Department of Geography, Research Paper 116.

New York Times. March 31, 1985.

Norris, R.E. et al. 1982. *Geography: An Introductory Perspective.* Columbus, OH: Merrill.

Ott, W.R. (editor). 1978. *Environmental Indices: Theory and Practice.* Ann Arbor, MI: Ann Arbor Science.

Pacione, M. 1982. "The Use of Objective and Subjective Measures of Life Quality in Human Geography," *Progress in Human Geography* 6:495-514.

Pierce, R.M. 1984. "Rating America's Cities: A Perceptual Analysis of Objective Measures," Paper presented at the Annual Meetings of the Association of American Geographers, Washington, DC.

Pierce, R.M. 1985. "Rating America's Metropolitan Areas," *American Demographics* 7 (7):20-25.

Porell, F.W. 1982. "Intermetropolitan Migration and Quality of Life," *Journal of Regional Science* 22:137-58.

Power, T.M. 1980. *The Economic Value of the Quality of Life.* Boulder, CO: Westview Press.

President's Commission for a National Agenda for the Eighties. 1980. *The Quality of American Life in the Eighties.* Washington, DC: Goverment Printing Office.

Ram, R. 1982. "Composite Indices of Physical Quality of Life, Basic Needs Fulfillment and Income," *Journal of Development Economics* 11:227-47.

Rau, J.G. and D.C. Wooten. 1980. *Environmental Impact Analysis Handbook.* New York: McGraw-Hill.

Rees, P.H. 1970. "Concepts of Social Space: Toward an Urban Social Geography," pp. 306-394 in B.J.L. Berry and F.S. Horton (editors), *Geographic Perspectives on Urban Systems.* Englewood Cliffs, NJ: Prentice-Hall.

Rees, P.H. 1971. "Factorial Ecology: An Extended Definition, Survey, and Critique of the Field," *Economic Geography* 47:220-233.

Rees, P.H. 1972. "Problems of Classifying Sub-areas Within Cities," in B.J.L. Berry (editor), *City Classification Handbook.* New York: Wiley.

Rees, P.H. 1979. *Residential Patterns in American Cities: 1960.* Chicago, IL: University of Chicago, Department of Geography, Research Paper 189.

Relph, E.C. 1976. *Place and Placelessness.* London: Pion.

Relph, E.C. 1981. *Rational Landscapes and Humanistic Geography.* London, UK: Croom Helm.

Renwick, H.L. and S.C. Cutter. 1983. "Wish You Were Here — Map Postcards and Images of Place," *Landscape* 27(1):30-38.

Roseman, C.C. 1976. "Migration of Whites to Central Cities and Quality of Life," *Geographical Survey* 5(1):14-21.

Rossi, R.J. and K.J. Gilmartin. 1980. *The Handbook of Social Indicators: Sources, Characteristics and Analysis.* New York: Garland STPM Press.

Rubenstein, C. 1982. "Regional States of Mind," *Psychology Today* February:22-30.

Saarinen, T.F. 1976. *Environmental Planning: Perception and Behavior.* Boston, MA: Houghton Mifflin.

Saarinen, T.F. and J.L. Sell. 1980. "Environmental Perception," *Progress in Human Geography* 4:535-548.

Saarinen, T.F. 1981. "Environmental Perception," *Progress in Human Geography* 5:525-547.

Saarinen, T.F. et al. (editors). 1984. *Environmental Perception and Behavior: An Inventory and Prospect.* Chicago: University of Chicago, Department of Geography, Research Paper 209.

Sell, J.L. et al. 1984. "Toward a Theoretical Framework for Landscape Perception," pp. 61-83 in Saarinen et al. 1984.

Sheldon, E.B. and W.E. Moore (editors). 1968. *Indicators of Social Change: Concepts and Measurements.* New York: Russell Sage Foundation.

Sheldon, E.B. and K.C. Land. 1972. "Social Reporting for the 1970s: A Review and Progammatic Statement," *Policy Sciences* 3(2):137-151.

Shevky, E. and W. Bell. 1955. *Social Area Analysis.* Stanford, CA: Stanford University Press.

Shonfield, A. and S. Shaw (editors). 1972. *Social Indicators and Social Policy.* London, UK: Heinemann.

Smith, D. 1973. *Geography of Social Well Being.* New York: McGraw Hill.

Smith, T.W. 1981. "Social Indicators: A Review Essay," *Journal of Social History* 14:739-47.

Suttles, G. 1968. *The Social Order of the Slum.* Chicago, IL: University of Chicago Press.

Suttles, G. 1972. *The Social Construction of Communities.* Chicago, IL: University of Chicago Press.

Szalai, A. and F.M. Andrews. 1981. *Quality of Life: Comparative Studies.* Beverly Hills, CA: Sage Studies in International Sociology.

Taylor, C.L. and D.A. Jodice. 1983. *World Handbook of Political and Social Indicators.* (3rd Edition). New Haven, CT: Yale University Press.

Thomas, W.A. (editor). 1972. *Indicators of Environmental Quality.* New York: Plenum.

Thorndike, E.L. 1939. *Your City.* New York: Harcourt, Brace and Co.

Thorndike, E.L. 1940. *144 Smaller Cities.* New York: Harcourt, Brace and Co.

Time Magazine. March 11, 1985 and April 8, 1985.

Tuan, Yi-Fu. 1974. *Topophilia: A Study of Environmental Perception, Attitudes and Values.* Englewood Cliffs, NJ: Prentice-Hall.

Tuan, Yi-Fu. 1977. *Space and Place.* Minneapolis, MN: University of Minnesota Press.

USA Today. February 28, 1985.

U.S. Bureau of Census. 1974. *Social Indicators.* Washington, DC: Government Printing Office.

U.S. Bureau of Census. 1977. *Social Indicators, 1976.* Washington, DC: Government Printing Office.

U.S. Bureau of Census. 1980. *Social Indicators III.* Washington, DC: Government Printing Office.

U.S. Bureau of Census. 1976-84. *Annual Housing Survey.* Washington, DC: Government Printing Office.

U.S. Council on Environmental Quality. 1982. *Environmental Quality 1982. 13th Annual Report.* Washington, DC: Government Printing Office.

U.S. Environmental Protection Agency. 1973a. *The Quality of Life Concept, A Potential New Tool for Decisionmakers.* Washington, DC: Government Printing Office.

U.S. Environmental Protection Agency. 1973b. *Studies in Environment, Volume II Quality of Life.* Washington, DC: Government Printing Office.

U.S. Department of Housing and Urban Development. 1978. *The 1978 HUD Survey on the Quality of Community Life: A Data Book.* Washington, DC: Government Printing Office.

Weinstein, N.D. 1976. "Human Evaluations of Environmental Noise," pp. 229-252 in Craik and Zube 1976.

Wilson, A.G. and M.J. Kirkby. 1980. *Mathematics for Geographers and Planners.* Oxford, UK: Clarendon Press.

Wingo, L. and A. Evans. 1977. *Public Economics and the Quality of Life.* Baltimore, MD: Johns Hopkins Press for Resources for the Future.

Wirth, L. 1928. *The Ghetto.* Chicago, IL: University of Chicago Press.

Zelinsky, W. 1980. "North America's Vernacular Regions," *Annals,* Association of American Geographers 70:1-16.

Zorbaugh, H. 1929. *The Gold Coast and the Slum.* Chicago, IL: University of Chicago Press.

Zube, E.H. *et al.* 1975. *Landscape Assessment: Values, Perceptions, and Resources.* Stroudsburg, PA: Dowden, Hutchinson, and Ross.